Battle Studies

by

Colonel Charles-Jean-Jacques-Joseph Ardant du Picq

The Echo Library 2006

Published by

The Echo Library

Echo Library
131 High St.
Teddington
Middlesex TW11 8HH

www.echo-library.com

Please report serious faults in the text to complaints@echo-library.com

ISBN 1-4068-3097-6

BATTLE STUDIES
ANCIENT AND MODERN BATTLE
BY COLONEL ARDANT DU PICQ
FRENCH ARMY

TRANSLATED FROM THE EIGHTH EDITION IN THE FRENCH BY
COLONEL JOHN N. GREELY
FIELD ARTILLERY, U.S. ARMY
AND MAJOR ROBERT C. COTTON
GENERAL STAFF (INFANTRY), U.S. ARMY
Joint Author of "Military Field Notebook"
1921

TRANSLATION OF A LETTER FROM MARSHAL FOCH TO MAJOR GENERAL A. W. GREELY, DATED MALSHERBE, OCTOBER 23, 1920

MY DEAR GENERAL:

Colonel Ardant du Picq was the exponent of *moral force*, the most powerful element in the strength of armies. He has shown it to be the preponderating influence in the outcome of battles.

Your son has accomplished a very valuable work in translating his writings. One finds his conclusions amply verified in the experience of the American Army during the last war, notably in the campaign of 1918.

Accept, my dear General, my best regards.
F. FOCH.

CONTENTS
PREFACE
TRANSLATOR'S NOTE
INTRODUCTION
A MILITARY THINKER
RECORD OF MILITARY SERVICE OF COLONEL ARDANT DU PICQ
EXTRACT FROM THE HISTORY OF THE 10TH INFANTRY
      REGIMENT

PART ONE: ANCIENT BATTLE
INTRODUCTION
I MAN IN PRIMITIVE AND ANCIENT COMBAT
II KNOWLEDGE OF MAN MADE ROMAN TACTICS; THE SUCCESSES
      OF HANNIBAL; THOSE OF CAESAR
III ANALYSIS OF THE BATTLE OF CANNAE
IV ANALYSIS OF THE BATTLE OF PHARSALUS AND SOME
      CHARACTERISTIC EXAMPLES
V MORALE IN ANCIENT BATTLE
VI HOW REAL COMBATANTS ARE OBTAINED AND HOW THE
      FIGHTING OF TO-DAY REQUIRES THEM TO BE MORE
      DEPENDABLE THAN IN ANCIENT BATTLE
VII PURPOSE OF THIS STUDY AND WHAT IS NECESSARY TO
      COMPLETE IT

PART TWO: MODERN BATTLE
I   GENERAL DISCUSSION
   1. Ancient and Modern Battle
   2. Moral Elements in Battle
   3. Material and Moral Effect
   4. The Theory of Strong Battalions
   5. Combat Methods
II INFANTRY
   1. Masses—Deep Columns
   2. Skirmishers—Supports—Reserves—Squares
   3. Firing
   4. Marches—Camps—Night Attacks
III CAVALRY
   1. Cavalry and Modern Appliances
   2. Cavalry Against Cavalry
   3. Cavalry Against Infantry
   4. Armor and Armament
IV ARTILLERY
V COMMAND, GENERAL STAFF AND ADMINISTRATION
VI SOCIAL AND MILITARY INSTITUTIONS; NATIONAL
      CHARACTERISTICS

APPENDICES
I MEMORANDUM ON INFANTRY FIRE

   1. Introduction
   2. Succinct History of the Development of Small Arms, from the Arquebus to Our Rifle
   3. Progressive Introduction of Fire-Arms Into the Armament of the Infantryman
   4. The Classes of Fire Employed with Each Weapon
   5. Methods of Fire Used in the Presence of the Enemy; Methods Recommended or Ordered but Impractical
   6. Fire at Will—Its Efficacy
   7. Fire by Rank Is a Fire to Occupy the Men in Ranks
   8. The Deadly Fire Is the Fire of Skirmishers
   9. The Absolute Impossibility of Fire at Command

II HISTORICAL DOCUMENTS

   1. Cavalry (An Extract from Xenophon)
   2. Marius Against the Cimbrians (Extract from Plutarch's "Life of Marius")
   3. The Battle of The Alma (Extract from the Correspondence of Colonel Ardant du Picq)
   4. The Battle of the Alma (Extract from the Correspondence of Colonel Ardant du Picq)
   5. The Battle of Inkermann (Extract from the Correspondence of Colonel Ardant du Picq)
   6. The Battle of Magenta (Extract from the Correspondence of Colonel Ardant du Picq)
   7. The Battle of Solferino (Extract from the Correspondence of Colonel Ardant du Picq)
   8. Mentana (Extract from the Correspondence of Colonel Ardant du Picq)

PREFACE

BY FRANK H. SIMONDS
Author of "History of the World War," "'They Shall Not Pass'—Verdun," Etc.

In presenting to the American reading public a translation of a volume written by an obscure French colonel, belonging to a defeated army, who fell on the eve of a battle which not alone gave France over to the enemy but disclosed a leadership so inapt as to awaken the suspicion of treason, one is faced by the inevitable interrogation—"Why?"

Yet the answer is simple. The value of the book of Ardant du Picq lies precisely in the fact that it contains not alone the unmistakable forecast of the defeat, itself, but a luminous statement of those fundamental principles, the neglect of which led to Gravelotte and Sedan.

Napoleon has said that in war the moral element is to all others as three is to one. Moreover, as du Picq impressively demonstrates, while all other circumstances change with time, the human element remains the same, capable of just so much endurance, sacrifice, effort, and no more. Thus, from Caesar to Foch, the essential factor in war endures unmodified.

And it is not the value of du Picq's book, as an explanation of the disasters of 1870, but of the triumphs of 1914-18, which gives it present and permanent interest. It is not as the forecast of why Bazaine, a type of all French commanders of the Franco-Prussian War, will fail, but why Foch, Joffre, Pétain will succeed, that the volume invites reading to-day.

Beyond all else, the arresting circumstances in the fragmentary pages, perfect in themselves but incomplete in the conception of their author, is the intellectual and the moral kinship they reveal between the soldier who fell just before the crowning humiliation of Gravelotte and the victor of Fère Champenoise, the Yser and the colossal conflict of 1918 to which historians have already applied the name of the Battle of France, rightly to suggest its magnitude.

Read the hastily compiled lectures of Foch, the teacher of the École de Guerre, recall the fugitive but impressive words of Foch, the soldier, uttered on the spur of the moment, filled with homely phrase, and piquant figure and underlying all, one encounters the same integral conception of war and of the relation of the moral to the physical, which fills the all too scanty pages of du Picq.

"For me as a soldier," writes du Picq, "the smallest detail caught on the spot and in the heat of action is more instructive than all the Thiers and the Jominis in the world." Compare this with Foch explaining to his friend André de Mariecourt, his own emotions at the critical hour at Fère Champenoise, when he had to invent something new to beguile soldiers who had retreated for weeks and been beaten for days. His tactical problem remained unchanged, but he must give his soldiers, tired with being beaten to the "old tune" a new air, which would appeal to them as new, something to which they had not been beaten, and the same philosophy appears.

Du Picq's contemporaries neglected his warning, they saw only the outward circumstances of the Napoleonic and Frederican successes. In vain du Picq warned them that the victories of Frederick were not the logical outgrowth of the minutiae of the Potsdam parades. But du Picq dead, the Third Empire fallen, France prostrated but not annihilated by the defeats of 1870, a new generation emerged, of which Foch was but the last and most shining example. And this generation went back, powerfully aided by the words of du Picq, to that older tradition, to the immutable principles of war.

With surprising exactness du Picq, speaking in the abstract, foretold an engagement in which the mistakes of the enemy would be counterbalanced by their energy in the face of French passivity, lack of any control conception. Forty years later in the École de Guerre, Foch explained the reasons why the strategy of Moltke, mistaken in all respects, failed to meet the ruin it deserved, only because at Gravelotte Bazaine could not make up his mind, solely because of the absence in French High Command of precisely that "Creed of Combat" the lack of which du Picq deplored.

Of the value of du Picq's work to the professional soldier, I naturally cannot speak, but even for the civilian, the student of military events, of war and of the larger as well as the smaller circumstances of battle, its usefulness can hardly be exaggerated. Reading it one understands something, at least of the soul as well as the science of combat, the great defeats and the great victories of history seem more intelligible in simple terms of human beings. Beyond this lies the contemporaneous value due to the fact that nowhere can one better understand Foch than through the reading of du Picq.

By translating this volume of du Picq and thus making it available for an American audience whose interest has been inevitably stirred by recent events, the translators have done a public as well as a professional service. Both officers enjoyed exceptional opportunities and experiences on the Western front. Col. Greely from Cantigny to the close of the battle of the Meuse-Argonne was not only frequently associated with the French army, but as Chief of Staff of our own First Division, gained a direct knowledge of the facts of battle, equal to that of du Picq, himself.

On the professional side the service is obvious, since before the last war the weakness of the American like the British Army, a weakness inevitable, given our isolation, lay in the absence of adequate study of the higher branches of military science and thus the absence of such a body of highly skilled professional soldiers, as constituted the French or German General Staff. The present volume is a clear evidence that American officers themselves have voluntarily undertaken to make good this lack.

On the non-professional side and for the general reader, the service is hardly less considerable, since it supplies the least technically informed with a simply comprehensible explanation of things which almost every one has struggled to grasp and visualize during the last six years extending from the battle of Marne in 1914 to that of the Vistula in 1920.

Of the truth of this latter assertion, a single example will perhaps suffice. Every forthcoming military study of the campaign of 1914 emphasizes with renewed energy the fact that underlying all the German conceptions of the opening operations was the purpose to repeat the achievement of Hannibal at Cannae, by bringing the French to battle under conditions which should, on a colossal scale, reproduce those of Hannibal's greatest victory. But nowhere better than in du Picq's volume, are set forth the essential circumstances of the combat which, after two thousand years gave to Field Marshal von Schlieffen the root ideas for the strategy expressed in the first six weeks of 1914. And, as a final observation, nowhere better than in du Picq's account, can one find the explanation of why the younger Moltke failed in executing those plans which gave Hannibal one of the most shining triumphs in all antiquity.

Thus, although he died in 1870, du Picq lives, through his book, as one of the most useful guides to a proper understanding of a war fought nearly half a century later.

FRANK H. SIMONDS.
Snowville, New Hampshire,
October 15, 1920.

TRANSLATORS' NOTE

Colonel Ardant du Picq's "Battle Studies" is a French military classic. It is known to every French army officer; it is referred to as an established authority in such works as Marshal Foch's "The Principles of War." It has been eagerly read in the original by such American army officers as have chanced upon it; probably only the scarcity of thinking men with military training has precluded the earlier appearance of an American edition.

The translators feel that the war with Germany which brought with it some military training for all the best brains of the country has prepared the field for an American edition of this book. They are sure that every American reader who has had actual battle experience in any capacity will at some point say to himself, "That is absolutely true...." or, "That reminds me of the day...."

Appendices II, III, IV, and V, appearing in the edition from which this translation is made, deal with issues and military questions entirely French and not of general application. They are therefore not considered as being of sufficient interest to be reproduced herein. Appendix VI of the original appears herein as Appendix II.

The translation is unpretentious. The translators are content to exhibit such a work to the American military public without changing its poignancy and originality. They hope that readers will enjoy it as much as they have themselves.
J. N. G.
R. C. C.

INTRODUCTION

We present to the public the complete works of Colonel Ardant du Picq, arranged according to the plan of the author, enlarged by unpublished fragments and documents.

These unpublished documents are partially known by those who have read "Studies on Combat" (Hachette & Dumaine, 1880). A second edition was called for after a considerable time. It has left ineffaceable traces in the minds of thinking men with experience. By its beauty and the vigor of its teachings, it has created in a faithful school of disciples a tradition of correct ideas.

For those familiar with the work, there is no need for emphasizing the importance and usefulness of this rejuvenated publication. In it they will find new sources of interest, which will confirm their admiration for the author.

They will also rejoice in the popularity of their teacher, already highly regarded in the eyes of his profession on account of his presentation of conclusions, the truth of which grows with years. His work merits widespread attention. It would be an error to leave it in the exclusive possession of special writers and military technicians. In language which is equal in power and pathetic beauty, it should carry its light much further and address itself to all readers who enjoy solid thought. Their ideas broadened, they will, without fail, join those already initiated.

No one can glance over these pages with indifference. No one can fail to be moved by the strong and substantial intellect they reveal. No one can fail to feel their profound depths. To facilitate treatment of a subject which presents certain difficulties, we shall confine ourselves to a succinct explanation of its essential elements, the general conception that unites them, and the purpose of the author. But we must not forget the dramatic mutilation of the work unfortunately never completed because of the glorious death of Ardant du Picq.

When Colonel Ardant du Picq was killed near Metz in 1870 by a Prussian shell, he left works that divide themselves into two well-defined categories:

(1) Completed works:
Pamphlet (printed in 1868 but not intended for sale), which forms the first part of the present edition: Ancient Battle.

A series of memoirs and studies written in 1865. These are partly reproduced in Appendices I and II herein.

(2) Notes jotted down on paper, sometimes developed into complete chapters not requiring additions or revision, but sometimes abridged and drawn up in haste. They reveal a brain completely filled with its subject, perpetually working, noting a trait in a rapid phrase, in a vibrating paragraph, in observations and recollections that a future revision was to compile, unite and complete.

The collection of these notes forms the second part: Modern Battle. These notes were inspired by certain studies or memoirs which are presented in Appendices I-V, and a Study on Combat, with which the Colonel was occupied, and of which we gave a sketch at the end of the pamphlet of 1868. He himself started research among the officers of his acquaintance, superiors, equals or subordinates, who had served in war. This occupied a great part of his life.

In order to collect from these officers, without change or misrepresentation, statements of their experiences while leading their men in battle or in their divers contacts with the enemy, he sent to each one a questionnaire, in the form of a circular. The reproduction herein is from the copy which was intended for General Lafont de Villiers, commanding the 21st Division at Limoges. It is impossible to over-emphasize the great value of this document which gives the key to the constant meditations of Ardant du Picq, the key to the reforms which his methodical and logical mind foresaw. It expounds a principle founded upon exact facts faithfully stated. His entire work, in embryo, can be seen between the lines of the questionnaire. This was his first attempt at reaction against the universal routine surrounding him.

From among the replies which he received and which his family carefully preserved, we have extracted the most conclusive. They will be found in Appendix II—Historical Documents. Brought to light, at the urgent request of the author, they complete the book, corroborating statements by examples. They illuminate his doctrines by authentic historical depositions.

In arranging this edition we are guided solely by the absolute respect which we have for the genius of Ardant du Picq. We have endeavored to reproduce his papers in their entirety, without removing or adding anything. Certain disconnected portions have an inspired and fiery touch which would be lessened by the superfluous finish of an attempt at editing. Some repetitions are to be found; they show that the appendices were the basis for the second part of the volume, Modern Battle. It may be stated that the work, suddenly halted in 1870, contains criticisms, on the staff for instance, which aim at radical reforms.

ERNEST JUDET.

BATTLE STUDIES

A MILITARY THINKER

Near Longeville-les-Metz on the morning of August 15, 1870, a stray projectile from a Prussian gun mortally wounded the Colonel of the 10th Regiment of the Line. The obscure gunner never knew that he had done away with one of the most intelligent officers of our army, one of the most forceful writers, one of the most clear-sighted philosophers whom sovereign genius had ever created.

Ardant du Picq, according to the Annual Register, commanded but a regiment. He was fitted for the first rank of the most exalted. He fell at the hour when France was thrown into frightful chaos, when all that he had foreseen, predicted and dreaded, was being terribly fulfilled. New ideas, of which he was the unknown trustee and unacknowledged prophet, triumphed then at our expense. The disaster that carried with it his sincere and revivifying spirit, left in the tomb of our decimated divisions an evidence of the necessity for reform. When our warlike institutions were perishing from the lack of thought, he represented in all its greatness the true type of military thinker. The virile thought of a military thinker alone brings forth successes and maintains victorious nations. Fatal indolence brought about the invasion, the loss of two provinces, the bog of moral miseries and social evils which beset vanquished States.

The heart and brain of Ardant du Picq guarded faithfully a worthy but discredited cult. Too frequently in the course of our history virtues are forsaken during long periods, when it seems that the entire race is hopelessly abased. The mass perceives too late in rare individuals certain wasted talents—treasures of sagacity, spiritual vigor, heroic and almost supernatural comprehension. Such men are prodigious exceptions in times of material decadence and mental laxness. They inherit all the qualities that have long since ceased to be current. They serve as examples and rallying points for other generations, more clear-sighted and less degenerate. On reading over the extraordinary work of Ardant du Picq, that brilliant star in the eclipse of our military faculties, I think of the fatal shot that carried him off before full use had been found for him, and I am struck by melancholy. Our fall appears more poignant. His premature end seems a punishment for his contemporaries, a bitter but just reproach.

Fortunately, more honored and believed in by his successors, his once unappreciated teaching contributes largely to the uplift and to the education of our officers. They will be inspired by his original views and the permanent virtue contained therein. They will learn therefrom the art of leading and training our young soldiers and can hope to retrieve the cruel losses of their predecessors.

Ardant du Picq amazes one by his tenacity and will power which, without the least support from the outside, animate him under the trying conditions of his period of isolated effort.

In an army in which most of the seniors disdained the future and neglected their responsibilities, rested satisfied on the laurels of former campaigns and relied on superannuated theories and the exercises of a poor parade, scorned foreign organizations and believed in an acquired and constant superiority that dispenses with all work, and did not suspect even the radical transformations which the development of rifles and rapid-fire artillery entail; Ardant du Picq worked for the common good. In his modest retreat, far from the pinnacles of glory, he tended a solitary shrine of unceasing activity and noble effort. He burned with the passions which ought to have moved the staff and higher commanders. He watched while his contemporaries slept.

Toward the existing system of instruction and preparation which the first blow shattered, his incorruptible honesty prevented him from being indulgent. While terrified leaders passed from arrogance or thoughtlessness to dejection and confusion, the blow was being struck. Served by his marvelous historical gifts, he studied the laws of ancient combat in the poorly interpreted but innumerable documents of the past. Then, guided by the immortal light which never failed, the feverish curiosity of this soldier's mind turned towards the research of the laws of modern combat, the subject of his preference. In this study he developed to perfection his psychological attainments. By the use of these attainments he simplified the theory of the conduct of war. By dissecting the motor nerves of the human heart, he released basic data on the essential principles of combat. He discovered the secret of combat, the way to victory.

Never for a second did Ardant du Picq forget that combat is the object, the cause of being, the supreme manifestation of armies. Every measure which departs therefrom, which relegates it to the middle ground is deceitful, chimerical, fatal. All the resources accumulated in time of peace, all the tactical evolutions, all the strategical calculations are but conveniences, drills, reference marks to lead up to it. His obsession was so overpowering that his presentation of it will last as long as history. This obsession is the rôle of man in combat. Man is the incomparable instrument whose elements, character, energies, sentiments, fears, desires, and instincts are stronger than all abstract rules, than all bookish theories. War is still more of an art than a science. The inspirations which reveal and mark the great strategists, the leaders of men, form the unforeseen element, the divine part. Generals of genius draw from the human heart ability to execute a surprising variety of movements which vary the routine; the mediocre ones, who have no eyes to read readily therein, are doomed to the worst errors.

Ardant du Picq, haunted by the need of a doctrine which would correct existing evils and disorders, was continually returning to the fountain-head. Anxious to instruct promising officers, to temper them by irrefutable lessons, to mature them more rapidly, to inspire them with his zeal for historical incidents, he resolved to carry on and add to his personal studies while aiding them. Daring to take a courageous offensive against the general inertia of the period, he translated the problem of his whole life into a series of basic questions. He presented in their most diverse aspects, the basic questions which perplex all military men, those of which knowledge in a varying degree of perfection distinguish and classify military men. The nervous grasp of an incomparable style models each of them, carves them with a certain harshness, communicates to them a fascinating yet unknown authority which crystallizes them in the mind, at the same time giving to them a positive form that remains true for all armies, for all past, present and future centuries. Herewith is the text of the concise and pressing questions which have not ceased to be as important to-day (1902) as they were in 1870:

"*General,*

"In the last century, after the improvements of the rifle and field artillery by Frederick, and the Prussian successes in war—to-day, after the improvement of the new rifle and cannon to which in part the recent victories are due—we find all thinking men in the army asking themselves the question: 'How shall we fight to-morrow?' We have no creed on the subject of combat. And the most opposing methods confuse the intelligence of military men.

"Why? A common error at the starting point. One might say that no one is willing to acknowledge that it is necessary to understand yesterday in order to know to-morrow, for the things of yesterday are nowhere plainly written. The lessons of yesterday exist solely in the memory of those who know how to remember because they have known how to see, and those individuals have never spoken. I make an appeal to one of those.

"The smallest detail, taken from an actual incident in war, is more instructive for me, a soldier, than all the Thiers and Jominis in the world. They speak, no doubt, for the heads of states and armies but they never show me what I wish to know—a battalion, a company, a squad, in action.

"Concerning a regiment, a battalion, a company, a squad, it is interesting to know: The disposition taken to meet the enemy or the order for the march toward them. What becomes of this disposition or this march order under the isolated or combined influences of accidents of the terrain and the approach of danger?

"Is this order changed or is it continued in force when approaching the enemy?

"What becomes of it upon arriving within the range of the guns, within the range of bullets?

"At what distance is a voluntary or an ordered disposition taken before starting operations for commencing fire, for charging, or both?

"How did the fight start? How about the firing? How did the men adapt themselves? (This may be learned from the results: So many bullets fired, so many men shot down—when such data are available.) How was the charge made? At what distance did the enemy flee before it? At what distance did the charge fall back before the fire or the good order and good dispositions of the enemy, or before such and such a movement of the enemy? What did it cost? What can be said about all these with reference to the enemy?

"The behavior, i.e., the order, the disorder, the shouts, the silence, the confusion, the calmness of the officers and men whether with us or with the enemy, before, during, and after the combat?

"How has the soldier been controlled and directed during the action? At what instant has he had a tendency to quit the line in order to remain behind or to rush ahead?

"At what moment, if the control were escaping from the leader's hands, has it no longer been possible to exercise it?

"At what instant has this control escaped from the battalion commander? When from the captain, the section leader, the squad leader? At what time, in short, if such a thing did take place, was there but a disordered impulse, whether to the front or to the rear carrying along pell-mell with it both the leaders and men?

"Where and when did the halt take place?

"Where and when were the leaders able to resume control of the men?

"At what moments before, during, or after the day, was the battalion roll-call, the company roll-call made? The results of these roll-calls?

"How many dead, how many wounded on the one side and on the other; the kind of wounds of the officers, non-commissioned officers, corporals, privates, etc., etc.?

"All these details, in a word, enlighten either the material or the moral side of the action, or enable it to be visualized. Possibly, a closer examination might show that they are matters infinitely more instructive to us as soldiers than all the discussions imaginable on the plans and general conduct of the campaigns of the greatest captain in the great movements of the battle field. From colonel to private we are soldiers, not generals, and it is therefore our trade that we desire to know.

"Certainly one cannot obtain all the details of the same incident. But from a series of true accounts there should emanate an ensemble of characteristic details which in themselves are very apt to show in a striking, irrefutable way what was necessarily and forcibly taking place at such and such a moment of an action in war. Take the estimate of the soldier obtained in this manner to serve as a base for what might possibly be a rational method of fighting. It will put us on guard against *a priori* and pedantic school methods.

"Whoever has seen, turns to a method based on his knowledge, his personal experience as a soldier. But experience is long and life is short. The experiences of each cannot therefore be completed except by those of others.

"And that is why, General, I venture to address myself to you for your experiences.

"Proofs have weight.

"As for the rest, whether it please you to aid or not, General, kindly accept the assurance of most respectful devotion from your obedient servant."

The reading of this unique document is sufficient to explain the glory that Ardant du Picq deserved. In no other career has a professional ever reflected more clearly the means of pushing his profession to perfection; in no profession has a deeper penetration of the resources been made.

It pleases me particularly to associate the two words 'penseur' and 'militaire,' which, at the present time, the ignorance of preconceived opinion too frequently separates. Because such opinion is on the verge of believing them to be incompatible and contradictory.

Yet no calling other than the true military profession is so fitted to excite brain activity. It is preëminently the calling of action, at the same time diverse in its combinations and changing according to the time and locality wherein it is put to practice. No other profession is more complex nor more difficult, since it has for its aim and reason the instruction of men to overcome by training and endurance the fatigue and perils against which the voice of self-preservation is raised in fear; in other words, to draw from nature what is most opposed and most antipathic to this nature.

There is, however, much of routine in the customs of military life, and, abuse of it may bring about gross satires which in turn bring it into derision. To be sure, the career has two phases because it must fulfill simultaneously two exigencies. From this persons of moderate capacity draw back and are horrified. They solve the question by the sacrifice of the one or the other. If one considers only the lower and somewhat vulgar aspect of military life it is found to be composed of monotonous obligations clothed in a mechanical procedure of indispensable repetition. If one learns to grasp it in its ensemble and large perspective, it will be found that the days of extreme trial demand prodigies of vigor, spirit, intelligence, and decision! Regarded from this angle and supported in this light, the commonplace things of wearisome garrison life have as counterweights certain sublime compensations. These compensations preclude the false and contemptible results which come from intellectual idleness and the habit of absolute submission. If it yields to their narcotic charms, the best brain grows rusty and atrophies in the long run. Incapable of virile labor, it rebels at a renewal of its processes in sane initiative. An army in which vigilance is not perpetual is sick until the enemy demonstrates it to be dead.

Far, then, from attaching routine as an indispensable companion to military discipline it must be shown continually that in it lies destruction and loss. Military discipline does not degenerate except when it has not known the cult of its vitality and the secret of its grandeur. The teachers of war have all placed this truth as a preface to their triumphs and we find the most illustrious teachers to be the most severe. Listen to this critique of Frederick the Great on the maneuvers which he conducted in Silesia:

"The great mistake in inspections is that you officers amuse yourselves with God knows what buffooneries and never dream in the least of serious service. This is a source of stupidity which would become most dangerous in case of a serious conflict. Take shoe-makers and tailors and make generals of them and they will not commit worse follies! These blunders are made on a small as well as on a large scale. Consequently, in the greatest number of regiments, the private is not well trained; in Zaramba's regiment he is the worst; in Thadden's he amounts to nothing; and to no more in Keller's, Erlach's, and Haager's. Why? Because the officers are lazy and try to get out of a difficulty by giving themselves the least trouble possible."

In default of exceptional generals who remold in some campaigns, with a superb stroke, the damaged or untempered military metal, it is of importance to supply it with the ideals of Ardant du Picq. Those who are formed by his image, by his book, will never fall into error. His book has not been written to please aesthetic preciseness, but with a sincerity which knows no limit. It therefore contains irrefutable facts and theories.

The solidity of these fragmentary pages defies time; the work interrupted by the German shell is none the less erected for eternity. The work has muscles, nerves and a soul. It has the transparent concentration of reality. A thought may be expressed by a single word. The terseness of the calcined phrase explains the interior fire of it all, the magnificent conviction of the author. The distinctness of outline, the most astounding brevity of touch, is such that the vision of the future bursts forth from the resurrection of the past. The work contains, indeed, substance and marrow of a prophetic experience.

Amidst the praise rendered to the scintillating beauties of this book, there is perhaps, none more impressive than that of Barbey d'Aurevilly, an illustrious literary man of a long and generous patrician lineage. His comment, kindled with lyric enthusiasm, is illuminating. It far surpasses the usual narrow conception of technical subjects. Confessing his professional ignorance in matters of war, his sincere eulogy of the eloquent amateur is therefore only the more irresistible.

"Never," writes Barbey d'Aurevilly, "has a man of action—of brutal action in the eyes of universal prejudice—more magnificently glorified the spirituality of war. Mechanics—abominable mechanics—takes possession of the world, crushing it under its stupid and irresistible wheels. By the action of newly discovered and improved appliances the science of war assumes vast proportions as a means of destruction. Yet here, amid the din of this upset modern world we find a brain sufficiently master of its own thoughts as not to permit itself to be dominated by these horrible discoveries which, we are told, would make impossible Fredericks of Prussia and Napoleons and lower them to the level of the private soldier! Colonel Ardant du Picq tells us somewhere that he has never had entire faith in the huge battalions which these two great men, themselves alone worth more than the largest battalions, believed in. Well, to-day, this vigorous brain believes no more in the mechanical or mathematical force which is going to abolish these great battalions. A calculator without the least emotion, who considers the mind of man the essential in war—because it is this mind that makes war—he surely sees better than anybody else a profound change in the exterior conditions of war which he must consider. But the spiritual conditions which are produced in war have not changed. Such, is the eternal mind of man raised to its highest power by discipline. Such, is the Roman cement of this discipline that makes of men indestructible walls. Such, is the cohesion, the solidarity between men and their leaders. Such, is the moral influence of the impulse which gives the certainty of victory.

"'To conquer is to advance,' de Maistre said one day, puzzled at this phenomenon of victory. The author of "Etudes sur le Combat" says more simply: 'To conquer is to be sure to overcome.' In fine, it is the mind that wins battles, that will always win them, that always has won them throughout the world's history. The spirituality, the moral quality of war, has not changed since those times. Mechanics, modern arms, all the artillery invented by man and his science, will not make an end to this thing, so lightly considered at the moment and called the human soul. Books like that of Ardant du Picq prevent it from being disdained. If no other effect should be produced by this sublime book, this one thing would justify it. But there will be others—do not doubt it—I wish merely to point out the sublimity of this didactic book which, for me, has wings like celestial poetry and which has carried me above and far away from the materialistic abjectness of my time. The technique of tactics and the science of war are beyond my province. I am not, like the author, erudite on maneuvers and the battle field. But despite my ignorance of things exclusively military, I have felt the truth of the imperious demonstrations with which it is replete, as one feels the presence of the sun behind a cloud. His book has over the reader that moral ascendancy which is everything in war and which determines success, according to the author. This ascendancy, like truth itself, is the sort which cannot be questioned. Coming from the superior mind of a leader who inspires faith it imposes obedience by its very strength. Colonel Ardant du Picq was a military writer only, with a style of his own. He has the Latin brevity and concentration. He retains his thought, assembles it and always puts it out in a compact phrase like a cartridge. His style has the rapidity and precision of the long-range arms which have dethroned the bayonet. He would have been a writer anywhere. He was a writer by nature. He was of that sacred phalanx of those who have a style all to themselves."

Barbey d'Aurevilly rebels against tedious technicalities. Carried away by the author's historical and philosophical faculties, he soars without difficulty to the plane of Ardant du Picq. In like manner, du Picq ranges easily from the most mediocre military operations to the analysis of the great functions of policy of government and the evolution of nations.

Who could have unraveled with greater finesse the causes of the insatiable desires of conquest by the new power which was so desirous of occupying the leading rôle on the world's stage? If our diplomats, our ministers and our generals had seized the warning of 1866, the date of the defeat of Austria, it is possible that we might have been spared our own defeats.

"Has an aristocracy any excuse for existing if it is not military? No. The Prussian aristocracy is essentially military. In its ranks it does accept officers of plebeian extraction, but only under condition that they permit themselves to be absorbed therein.

"Is not an aristocracy essentially proud? If it were not proud it would lack confidence. The Prussian aristocracy is, therefore, haughty; it desires domination by force and its desire to rule, to dominate more and more, is the essence of its existence. It rules by war; it wishes war; it must have war at the proper time. Its leaders have the good judgment to choose the right moment. This love of war is in the very fiber, the very makeup of its life as an aristocracy.

"Every nation that has an aristocracy, a military nobility, is organized in a military way. The Prussian officer is an accomplished gentleman and nobleman; by instruction or examination he is most capable; by education, most worthy. He is an officer and commands from two motives, the French officer from one alone.

"Prussia, in spite of all the veils concealing reality, is a military organization conducted by a military corporation. A nation, democratically constituted, is not organized from a military point of view. It is, therefore, as against the other, in a state of unpreparedness for war.

"A military nation and a warlike nation are not necessarily the same. The French are warlike from organization and instinct. They are every day becoming less and less military.

"In being the neighbor of a military nation, there is no security for a democratic nation; the two are born enemies; the one continually menaces the good influences, if not the very existence of the other. As long as Prussia is not democratic she is a menace to us.

"The future seems to belong to democracy, but, before this future is attained by Europe, who will say that victory and domination will not belong for a time to military organization? It will presently perish for the lack of sustenance of life, when having no more foreign enemies to vanquish, to watch, to fight for control, it will have no reason for existence."

In tracing a portrait so much resembling bellicose and conquering Prussia, the sharp eye of Ardant du Picq had recognized clearly the danger which immediately threatened us and which his deluded and trifling fellow citizens did not even suspect. The morning after Sadowa, not a single statesman or publicist had yet divined what the Colonel of the 10th Regiment of the Line had, at first sight, understood. Written before the catastrophes of Froeschwiller, Metz and Sedan, the fragment seems, in a retrospective way, an implacable accusation against those who deceived themselves about the Hohenzollern country by false liberalism or a softening of the brain.

Unswerved by popular ideas, by the artificial, by the trifles of treaties, by the chimera of theories, by the charlatanism of bulletins, by the nonsense of romantic fiction, by the sentimentalities of vain chivalry, Ardant du Picq, triumphant in history, is even more the incomparable master in the field of his laborious days and nights, the field of war itself. Never has a clearer vision fathomed the bloody mysteries of the formidable test of war. Here man appears as his naked self. He is a poor thing when he succumbs to unworthy deeds and panics. He is great under the impulse of voluntary sacrifice which transforms him under fire and for honor or the salvation of others makes him face death.

The sound and complete discussions of Ardant du Picq take up, in a poignant way, the setting of every military drama. They envelop in a circle of invariable phenomena the apparent irregularity of combat, determining the critical point in the outcome of the battle. Whatever be the conditions, time or people, he gives a code of rules which will not perish. With the enthusiasm of Pascal, who should have been a soldier, Ardant du Picq has the preëminent gift of expressing the infinite in magic words. He unceasingly opens an abyss under the feet of the reader. The whole metaphysics of war is contained therein and is grasped at a single glance.

He shows, weighed in the scales of an amazing exactitude, the normal efficiency of an army; a multitude of beings shaken by the most contradictory passions, first desiring to save their own skins and yet resigned to any risk for the sake of a principle. He shows the quantity and quality of possible efforts, the aggregate of losses, the effects of training and impulse, the intrinsic value of the troops engaged. This value is the sum of all that the leader can extract from any and every combination of physical preparation, confidence, fear of punishment, emulation, enthusiasm, inclination, the promise of success, administration of camps, fire discipline, the influence of ability and superiority, etc. He shows the tragic depths, so somber below, so luminous above, which appear in the heart of the combatant torn between fear and duty. In the private soldier the sense of duty may spring from blind obedience; in the non-commissioned officer, responsible for his detachment, from devotion to his trade; in the commanding officer, from supreme responsibility! It is in battle that a military organization justifies its existence. Money spent by the billions, men trained by the millions, are gambled on one irrevocable moment. Organization decides the terrible contest which means the triumph or the downfall of the nation! The harsh rays of glory beam above the field of carnage, destroying the vanquished without scorching the victor.

Such are the basic elements of strategy and tactics!

There is danger in theoretical speculation of battle, in prejudice, in false reasoning, in pride, in braggadocio. There is one safe resource, the return to nature.

The strategy that moves in elevated spheres is in danger of being lost in the clouds. It becomes ridiculous as soon as it ceases to conform to actual working tactics. In his classical work on the decisive battle of August 18, 1870, Captain Fritz Hoenig has reached a sound conclusion. After his biting criticism of the many gross errors of Steinmetz and Zastrow, after his description of the triple panic of the German troops opposite the French left in the valley and the ravine of the Mance, he ends by a reflection which serves as a striking ending to the book. He says, "The grandest illustration of Moltke's strategy was the battle of Gravelotte-Saint Privat; but the battle of Gravelotte has taught us one thing, and that is, the best strategy cannot produce good results if tactics is at fault."

The right kind of tactics is not improvised. It asserts itself in the presence of the enemy but it is learned before meeting the enemy.

"There are men," says Ardant du Picq, "such as Marshal Bugeaud, who are born military in character, mind, intelligence and temperament. Not all leaders are of this stamp. There is, then, need for standard or regulation tactics appropriate to the national character which should be the guide for the ordinary commander and which do not exact of him the exceptional qualities of a Bugeaud."

"Tactics is an art based on the knowledge of how to make men fight with their maximum energy against fear, a maximum which organization alone can give."

"And here confidence appears. It is not the enthusiastic and thoughtless confidence of tumultuous or improvised armies that gives way on the approach of danger to a contrary sentiment which sees treason everywhere; but the intimate, firm, conscious confidence which alone makes true soldiers and does not disappear at the moment of action."

"We now have an army. It is not difficult for us to see that people animated by passions, even people who know how to die without flinching, strong in the face of death, but without discipline and solid organization, are conquered by others who are individually less valiant but firmly organized, all together and one for all."

"Solidarity and confidence cannot be improvised. They can be born only of mutual acquaintanceship which establishes pride and makes unity. And, from unity comes in turn the feeling of force, that force which gives to the attack the courage and confidence of victory. Courage, that is to say, the domination of the will over instinct even in the greatest danger, leads finally to victory or defeat."

In asking for a doctrine in combat and in seeking to base it on the moral element, Ardant du Picq was ahead of his generation. He has had a very great influence. But, the doctrine is not yet established.

How to approach the adversary? How to pass from the defensive to the offensive? How to regulate the shock? How to give orders that can be executed? How to transmit them surely? How to execute them by economizing precious lives? Such are the distressing problems that beset generals and others in authority. The result is that presidents, kings and emperors hesitate, tremble, interrogate, pile reports upon reports, maneuvers upon maneuvers, retard the improvement of their military material, their organization, their equipment.

The only leaders who are equal to the difficulties of future war, come to conclusions expressed in almost the same terms. Recently General de Negrier, after having insisted that physical exhaustion determined by the nervous tension of the soldier, increased in surprising proportions according to the invisibility of the adversary, expressed himself as follows:

"The tide of battle is in the hands of each fighter, and never, at any time, has the individual bravery of the soldier had more importance.

"Whatever the science of the superior commander, the genius of his strategic combinations, the precision of his concentrations, whatever numerical superiority he may have, victory will escape him if the soldier does not conduct himself without being watched, and if he is not personally animated by the resolution to conquer or to perish. He needs much greater energy that formerly.

"He no longer has the intoxication of ancient attacks in mass to sustain him. Formerly, the terrible anxiety of waiting made him wish for the violent blow, dangerous, but soon passed. Now, all his normal and physical powers are tried for long hours and, in such a test, he will have but the resoluteness of his own heart to sustain him.

"Armies of to-day gain decisions by action in open order, where each soldier must act individually with will and initiative to attack the enemy and destroy him.

"The Frenchman has always been an excellent rifleman, intelligent, adroit and bold. He is naturally brave. The metal is good; the problem is to temper it. It must be recognized that to-day this task is not easy. The desire for physical comfort, the international theories which come therefrom, preferring economic slavery and work for the profit of the stranger to the struggle, do not incite the Frenchman to give his life in order to save that of his brother.

"The new arms are almost valueless in the hands of weakhearted soldiers, no matter what their number may be. On the contrary, the demoralizing power of rapid and smokeless firing, which certain armies still persist in not acknowledging, manifests itself with so much the more force as each soldier possesses greater valor and cool energy.

"It is then essential to work for the development of the moral forces of the nation. They alone will sustain the soldier in the distressing test of battle where death comes unseen.

"That is the most important of the lessons of the South African war. Small nations will find therein the proof that, in preparing their youth for their duties as soldiers and creating in the hearts of all the wish for sacrifice, they are certain to live free; but only at this price."

This profession of faith contradicts the imbecile sophisms foolishly put into circulation by high authority and a thoughtless press, on the efficiency of the mass, which is nothing but numbers, on the fantastic value of new arms, which are declared sufficient for gaining a victory by simple mechanical perfection, on the suppression of individual courage. It is almost as though courage had become a superfluous and embarrassing factor. Nothing is more likely to poison the army. Ardant du Picq is the best specific against the heresies and the follies of ignorance or of pedantry. Here are some phrases of unerring truth. They ought to be impressed upon all memories, inscribed upon the walls of our military schools. They ought to be learned as lessons by our officers and they ought to rule them as regulations and pass into their blood:

"Man is capable of but a given quantity of fear. To-day one must swallow in five minutes the dose that one took in an hour in Turenne's day."

"To-day there is greater need than ever for rigid formation."

"Who can say that he never felt fear in battle? And with modern appliances, with their terrible effect on the nervous system, discipline is all the more necessary because one fights only in open formation."

"Combat exacts a moral cohesion, a solidarity more compact that ever before."

"Since the invention of fire arms, the musket, rifle, cannon, the distances of mutual aid and support are increased between the various arms. The more men think themselves isolated, the more need they have of high morale."

"We are brought by dispersion to the need of a cohesion greater than ever before."

"It is a truth, so clear as to be almost naïve, that if one does not wish bonds broken, he should make them elastic and thereby strengthen them."

"It is not wise to lead eighty thousand men upon the battle field, of whom but fifty thousand will fight. It would be better to have fifty thousand all of whom would fight. These fifty thousand would have their hearts in the work more than the others, who should have confidence in their comrades but cannot when one-third of them shirk their work."

"The rôle of the skirmisher becomes more and more predominant. It is more necessary to watch over and direct him as he is used against deadlier weapons and as he is consequently more prone to try to escape from them at all costs in any direction."

"The thing is then to find a method that partially regulates the action of our soldiers who advance by fleeing or escape by advancing, as you like, and if something unexpected surprises them, escape as quickly by falling back."

"Esprit de corps improves with experience in wars. War becomes shorter and shorter, and more and more violent; therefore, create in advance an esprit de corps."

These truths are eternal. This whole volume is but their masterful development. They prove that together with audacious sincerity in the coördination of facts and an infallible judgment, Ardant du Picq possessed prescience in the highest degree. His prophetic eye distinguished sixty years ago the constituent principles of a good army. These are the principles which lead to victory. They are radically opposed to those which enchant our parliamentarians or military politicians, which are based on a fatal favoritism and which precipitate wars.

Ardant du Picq is not alone a superior doctrinaire. He will be consulted with profit in practical warlike organization. No one has better depicted the character of modern armies. No one knew better the value of what Clausewitz called, "The product of armed force and the country's force ... the heart and soul of a nation."

No more let us forget that he launched, before the famous prediction of von der Goltz, this optimistic view well calculated to rekindle the zeal of generals who struggle under the weight of enormous tasks incident to obligatory service.

"Extremes meet in many things. In the ancient times of conflict with pike and sword, armies were seen to conquer other solid armies even though one against two. Who knows if the perfection of long-range arms might not bring back these heroic victories? Who knows whether a smaller number by some combination of good sense or genius, or morale, and of appliances will not overcome a greater number equally well armed?"

After the abandonment of the law of 1872, and the repeal of the law of 1889, and before the introduction of numerous and disquieting reforms in recruitment and consequently, in the education of our regiments, would it not be opportune to study Ardant du Picq and look for the secret of force in his ideas rather than in the deceptive illusions of military automatism and materialism?

The martial mission of France is no more ended than war itself. The severities of war may be deplored, but the precarious justice of arbitration tribunals, still weak and divested of sanction, has not done away with its intervention in earthly quarrels. I do not suppose that my country is willing to submit to the mean estate, scourged with superb contempt by Donoso Cortes, who says:—

"When a nation shows a civilized horror of war, it receives directly the punishment of its mistake. God changes its sex, despoils it of its common mark of virility, changes it into a feminine nation and sends conquerors to ravish it of its honor."

France submits sometimes to the yoke of subtle dialecticians who preach total disarmament, who spread insanely disastrous doctrine of capitulation, glorify disgrace and humiliation, and stupidly drive us on to suicide. The manly counsels of Ardant du Picq are admirable lessons for a nation awakening. Since she must, sooner or later, take up her idle sword again, may France learn from him to fight well, for herself and for humanity!
ERNEST JUDET.
PARIS, October 10, 1902.

Ardant du Picq has said little about himself in his writings. He veils with care his personality. His life and career, little known, are the more worthy of the reader's interest, because the man is as original as the writer. To satisfy a natural curiosity, I asked the Colonel's family for the details of his life, enshrined in their memory. His brother has kindly furnished them in a letter to me. It contains many unpublished details and shows traits of character which confirm our estimate of the man, Ardant du Picq. It completes very happily the impression made by his book.

"PARIS, October 12, 1903.

"*Sir,*

"Herewith are some random biographical notes on the author of 'Etudes sur le Combat' which you requested of me.

"My brother entered Saint-Cyr quite late, at twenty-one years, which was I believe the age limit at that time. This was not his initial preference. He had a marked preference for a naval career, in which adventure seemed to offer an opportunity for his activity, and which he would have entered if the circumstances had so permitted. His childhood was turbulent and somewhat intractable; but, attaining adolescence, he retained from his former violence a very pronounced taste for physical exercise, especially for gymnastics, little practiced then, to which he was naturally inclined by his agility and muscular strength.

"He was successful in his classes, very much so in studies which were to his taste, principally French composition. In this he rose above the usual level of schoolboy exercises when the subject interested him. Certain other branches that were uninteresting or distasteful to him, as for instance Latin Grammar, he neglected. I do not remember ever having seen him attend a distribution of prizes, although he was highly interested, perhaps because he was too interested. On these occasions, he would disappear generally after breakfast and not be seen until evening. His bent was toward mechanical notions and handiwork. He was not uninterested in mathematics but his interest in this was ordinary. He was nearly refused entrance to Saint-Cyr. He became confused before the examiners and the results of the first part of the tests were almost negligible. He consoled himself with his favorite maxim as a young man: 'Onward philosophy.' Considering the first test as over and done with, he faced the second test with perfect indifference. This attitude gave him another opportunity and he came out with honors. As he had done well with the written test on 'Hannibal's Campaigns,' he was given a passing grade.

"At school he was liked by all his comrades for his good humor and frank and sympathetic character. Later, in the regiment, he gained naturally and without effort the affection of his equals and the respect of his subordinates. The latter were grateful to him for the real, cordial and inspiring interest he showed in their welfare, for he was familiar with the details of the service and with the soldier's equipment. He would not compromise on such matters and prevaricators who had to do with him did not emerge creditably.

"It can be said that after reaching manhood he never lied. The absolute frankness from which he never departed under any circumstances gave him prestige superior to his rank. A mere Lieutenant, he voted 'No' to the Coup d'Etat of December 2, and was admonished by his colonel who was sorry to see him compromise thus his future. He replied with his usual rectitude: 'Colonel, since my opinion was asked for, I must suppose that it was wanted.'

"On the eve of the Crimean war, his regiment, (67th) not seeming destined to take the field, he asked for and obtained a transfer to the light infantry (9th Battalion). It was with this battalion that he served in the campaign. When it commenced, he made his first appearance in the fatal Dobrutscha expedition. This was undertaken in a most unhealthy region, on the chance of finding there Cossacks who would have furnished matter for a communiqué. No Cossacks were found, but the cholera was. It cut down in a few hours, so as to speak, a large portion of the total strength. My brother, left with the rear guard to bury the dead, burn their effects and bring up the sick, was in his turn infected. The attack was very violent and he recovered only because he would not give in to the illness. Evacuated to the Varna hospital, he was driven out the first night by the burning of the town and was obliged to take refuge in the surrounding fields where the healthfulness of the air gave him unexpected relief. Returned to France as a convalescent, he remained there until the month of December (1854). He then rejoined his regiment and withstood to the end the rigors of the winter and the slowness of the siege.

"Salle's division to which the Trochu brigade belonged, and in which my brother served, was charged with the attack on the central bastion. This operation was considered a simple diversion without a chance of success. My brother, commanding the storming column of his battalion, had the good fortune to come out safe and sound from the deadly fire to which he was exposed and which deprived the battalion of several good officers. He entered the bastion with a dozen men. All were naturally made prisoners after a resistance which would have cost my brother his life if the bugler at his side had not warded off a saber blow at his head. Upon his return from captivity, in the first months of 1856, he was immediately made major in the 100th Regiment of the Line, at the instance of General Trochu who regarded him highly. He was called the following year to the command of the 16th Battalion of Foot Chasseurs. He served with this battalion during the Syrian campaign where there was but little serious action.

"Back again in France, his promotion to the grade of lieutenant-colonel, notwithstanding his excellent ratings and his place on the promotion list, was long retarded by the ill-will of Marshal Randon, the Minister of War. Marshal Randon complained of his independent character and bore him malice from an incident relative to the furnishing of shoes intended for his battalion. My brother, questioned by Marshal Niel about the quality of the lot of shoes, had frankly declared it bad.

"Promoted finally to lieutenant-colonel in the 55th in Algeria, he took the field there in two campaigns, I believe. Appointed colonel of the 10th of the Line in February, 1869, he was stationed at Lorient and at Limoges during the eighteen months before the war with Germany. He busied himself during this period with the preparation of his work, soliciting from all sides first-hand information. It was slow in coming in, due certainly to indifference rather than ill-will. He made several trips to Paris for the purpose of opening the eyes of those in authority to the defective state of the army and the perils of the situation. Vain attempts! 'They take all that philosophically,' he used to say.

"Please accept, Sir, with renewed acknowledgements of gratitude, the expression of my most distinguished sentiments.

"C. ARDANT DU PICQ.

"P. S. As to the question of atavism in which you showed some interest in our first conversation, I may say that our paternal line does not in my knowledge include any military man. The oldest ancestor I know of, according to an album of engravings by Albert Dürer, recovered in a garret, was a gold and silversmith at Limoges towards the end of the sixteenth century. His descendants have always been traders down to my grandfather who, from what I have heard said, did not in the least attend to his trade. The case is different with my mother's family which came from Lorraine. Our great-grandfather was a soldier, our grandfather also, and two, at least, of my mother's brothers gave their lives on the battlefields of the First Empire. At present, the family has two representatives in the army, the one a son of my brother's, the other a first cousin, once removed, both bearing our name.

"C. A. DU P."

RECORD OF MILITARY SERVICE OF COLONEL ARDANT DU PICQ

Ardant du Picq (Charles-Jean-Jacques-Joseph), was born October 19, 1821 at Périgueux (Dordogne). Entered the service as a student of the Special Military School, November 15, 1842.
Sub-Lieutenant in the 67th Regiment of the Line, October 1, 1844.
Lieutenant, May 15, 1848.
Captain, August 15, 1852.
Transferred to the 9th Battalion of Foot Chasseurs, December 25, 1853.
Major of the 100th Regiment of the Line, February 15, 1856.
Transferred to the 16th Battalion of Chasseurs, March 17, 1856.
Transferred to the 37th Regiment of the Line, January 23, 1863.
Lieutenant Colonel of the 55th Regiment of the Line, January 16, 1864.
Colonel of the 10th Regiment of Infantry of the Line, February 27, 1869.
Died from wounds at the military hospital in Metz, August 18, 1870.

CAMPAIGNS AND WOUNDS

Orient, March 29, 1854 to May 27, 1856. Was taken prisoner of war at the storming of the central bastion (Sebastopol) September 8, 1855; returned from enemy's prisons December 13, 1855.
Served in the Syrian campaign from August 6, 1860 to June 18, 1861; in Africa from February 24, 1864 to April 14, 1866; in Franco-German war, from July 15, 1870 to August 18, 1870.
Wounded—a comminute fracture of the right thigh, a torn gash in the left thigh, contusion of the abdomen—by the bursting of a projectile, August 15, 1870, Longeville-les-Metz (Moselle).

DECORATIONS

Chevalier of the Imperial Order of the Legion of Honor, Dec. 29, 1860.
Officer of the Imperial Order of the Legion of Honor, September 10, 1868.
Received the medal of H. M. the Queen of England.
Received the medal for bravery in Sardinia.
Authorized to wear the decoration of the fourth class of the Ottoman Medjidie order.

EXTRACT FROM THE HISTORY OF THE 10TH INFANTRY REGIMENT

CAMPAIGN OF 1870

On the 22nd of July, the three active battalions of the 10th Regiment of Infantry of the Line left Limoges and Angoulême by rail arriving on the 23rd at the camp at Châlons, where the 6th Corps of the Rhine Army was concentrating and organizing, under the command of Marshal Canrobert. The regiment, within this army corps, belonged to the 1st Brigade (Pechot) of the 1st Division (Tixier).

The organization on a war footing of the 10th Regiment of Infantry of the Line, begun at Limoges, was completed at the Châlons camp.

The battalions were brought up to seven hundred and twenty men, and the regiment counted twenty-two hundred and ten present, not including the band, the sappers and the headquarters section, which raised the effectives to twenty-three hundred men.

The troops of the 6th Corps were soon organized and Marshal Canrobert reviewed them on the 31st of July.

On August 5th, the division received orders to move to Nancy. It was placed on nine trains, of which the first left at 6 A. M. Arriving in the evening at its destination, the 1st brigade camped on the Leopold Racetrack, and the 10th Regiment established itself on the Place de la Grève.

The defeats of Forbach and Reichshofen soon caused these first plans to be modified. The 6th Corps was ordered to return to the Châlons camp. The last troops of the 2d Brigade, held up at Toul and Commercy, were returned on the same trains.

The 1st Brigade entrained at Nancy, on the night of August 8th, arriving at the Châlons camp on the afternoon of August 8th.

The 6th Corps, however, was to remain but a few days in camp. On the 10th it received orders to go to Metz. On the morning of the 11th the regiment was again placed on three successive trains. The first train carrying the staff and the 1st Battalion, arrived at Metz without incident. The second train, transporting the 2d Battalion and four companies of the 3d was stopped at about 11 P.M. near the Frouard branch.

The telegraph line was cut by a Prussian party near Dieulouard, for a length of two kilometers, and it was feared the road was damaged.

In order not to delay his arrival at Metz, nor the progress of the trains following, Major Morin at the head of the column, directed his commands to detrain and continue to Metz.

He caused the company at the head of the train to alight (6th Company, 2d Battalion, commanded by Captain Valpajola) and sent it reconnoitering on the road, about three hundred meters in advance of the train. All precautions were taken to assure the security of the train, which regulated its progress on that of the scouts.

After a run of about eight kilometers in this way, at Marbache station, all danger having disappeared and communication with Metz having been established, the train resumed its regulation speed. In consequence of the slowing up of the second column, the third followed at a short distance until it also arrived. On the afternoon of the 12th, the regiment was entirely united.

The division of which it was a part was sent beyond Montigny and it camped there as follows:

The 9th Chasseurs and 4th Regiment of the Line, ahead of the Thionville railroad, the right on the Moselle, the left on the Pont-à-Mousson highway; the 10th Regiment of the Line, the right supported at the branch of the Thionville and Nancy lines, the left in the direction of Saint-Privat, in front of the Montigny repair shops of the Eastern Railroad lines.

The regiment was thus placed in the rear of a redoubt under construction. The company of engineers was placed at the left of the 10th near the earthworks on which it was to work.

Along the ridge of the plateau, toward the Seille, was the 2d Brigade, which rested its left on the river and its right perpendicular to the Saint-Privat road, in rear of the field-work of this name. The divisional batteries were behind it.

The division kept this position August 13th and during the morning of the 14th. In the afternoon, an alarm made the division take arms, during the engagement that took place on the side of Vallières and Saint-Julien (battle of Borny). The regiment immediately occupied positions on the left of the village of Montigny.

At nightfall, the division retired to the rear of the railroad cut, and received orders to hold itself in readiness to leave during the night.

The regiment remained thus under arms, the 3d Battalion (Major Deschesnes), passing the night on grand guard in front of the Montigny redoubt.

Before daybreak, the division marched over the bank of the Thionville railroad, crossed the Moselle, and, marching towards Gravelotte, descended into the plain south of Longeville-les-Metz, where the principal halt was made and coffee prepared.

Scarcely had stacks been made, and the men set to making fires, about 7 A.M. when shells exploded in the midst of the troops. The shots came from the Bradin farm, situated on the heights of Montigny, which the division had just left the same morning, and which a German cavalry reconnaissance patrol supported by two pieces had suddenly occupied.

The Colonel had arms taken at once and disposed the regiment north of the road which, being elevated, provided sufficient cover for defilading the men.

He himself, stood in the road to put heart into his troops by his attitude, they having been a little startled by this surprise and the baptism of fire which they received under such disadvantageous circumstances.

Suddenly, a shell burst over the road, a few feet from the Colonel, and mutilated his legs in a frightful manner.

The same shell caused other ravages in the ranks of the 10th. The commander of the 3d Battalion, Major Deschesnes, was mortally wounded,

Captain Reboulet was killed, Lieutenant Pone (3d Battalion, 1st Company), and eight men of the regiment were wounded. The Colonel was immediately taken to the other side of the highway into the midst of his soldiers and a surgeon called, those of the regiment being already engaged in caring for the other victims of the terrible shot.

In the meantime, Colonel Ardant du Picq asked for Lieut.-Colonel Doleac, delivered to him his saddlebags containing important papers concerning the regiment and gave him his field glasses. Then, without uttering the least sound of pain, notwithstanding the frightful injury from which he must have suffered horribly, he said with calmness: "My regret is to be struck in this way, without having been able to lead my regiment on the enemy."

They wanted him to take a little brandy, he refused and accepted some water which a soldier offered him.

A surgeon arrived finally. The Colonel, showing him his right leg open in two places, made with his hand the sign of amputating at the thigh, saying: "Doctor, it is necessary to amputate my leg here."

At this moment, a soldier wounded in the shoulder, and placed near the Colonel, groaned aloud. Forgetting his own condition, the Colonel said immediately to the surgeon: "See first, doctor, what is the matter with this brave man; I can wait."

Because of the lack of instruments it was not possible to perform the amputation on the ground, as the Colonel desired, so this much deplored commander was transported to the Metz hospital.

Four days later (19th of August), Colonel Ardant du Picq died like a hero of old, without uttering the least complaint. Far from his regiment, far from his family, he uttered several times the words which summed up his affections: "My wife, my children, my regiment, adieu!"

PART ONE

ANCIENT BATTLE

INTRODUCTION

Battle is the final objective of armies and man is the fundamental instrument in battle. Nothing can wisely be prescribed in an army—its personnel, organization, discipline and tactics, things which are connected like the fingers of a hand—without exact knowledge of the fundamental instrument, man, and his state of mind, his morale, at the instant of combat.

It often happens that those who discuss war, taking the weapon for the starting point, assume unhesitatingly that the man called to serve it will always use it as contemplated and ordered by the regulations. But such a being, throwing off his variable nature to become an impassive pawn, an abstract unit in the combinations of battle, is a creature born of the musings of the library, and not a real man. Man is flesh and blood; he is body and soul. And, strong as the soul often is, it can not dominate the body to the point where there will not be a revolt of the flesh and mental perturbation in the face of destruction.

The human heart, to quote Marshal de Saxe, is then the starting point in all matters pertaining to war.

Let us study the heart, not in modern battle, complicated and not readily grasped, but in ancient battle. For, although nowhere explained in detail, ancient battle was simple and clear.

Centuries have not changed human nature. Passions, instincts, among them the most powerful one of self-preservation, may be manifested in various ways according to the time, the place, the character and temperament of the race. Thus in our times we can admire, under the same conditions of danger, emotion and anguish, the calmness of the English, the dash of the French, and that inertia of the Russians which is called tenacity. But at bottom there is always found the same man. It is this man that we see disposed of by the experts, by the masters, when they organize and discipline, when they order detailed combat methods and take general dispositions for action. The best masters are those who know man best, the man of today and the man of history. This knowledge naturally comes from a study of formations and achievements in ancient war.

The development of this work leads us to make such an analysis, and from a study of combat we may learn to know man.

Let us go even back of ancient battle, to primeval struggle. In progressing from the savage to our times we shall get a better grasp of life.

And shall we then know as much as the masters? No more than one is a painter by having seen the methods of painting. But we shall better understand these able men and the great examples they have left behind them.

We shall learn from them to distrust mathematics and material dynamics as applied to battle principles. We shall learn to beware of the illusions drawn from the range and the maneuver field.

There, experience is with the calm, settled, unfatigued, attentive, obedient soldier, with an intelligent and tractable man-instrument in short, and not with the nervous, easily swayed, moved, troubled, distrait, excited, restless being, not even under self-control, who is the fighting man from general to private. There are strong men, exceptions, but they are rare.

These illusions, nevertheless, stubborn and persistent, always repair the very next day the most damaging injuries inflicted on them by experience. Their least dangerous effect is to lead to prescribing the impractical, as if ordering the impractical were not really an attack on discipline, and did not result in disconcerting officers and men by the unexpected and by surprise at the contrast between battle and the theories of peacetime training.

Battle, of course, always furnishes surprises. But it furnishes less in proportion as good sense and the recognition of truth have had their effect on the training of the fighting man, and are disseminated in the ranks. Let us then study man in battle, for it is he who really fights.

CHAPTER I

MAN IN PRIMITIVE AND ANCIENT COMBAT

Man does not enter battle to fight, but for victory. He does everything that he can to avoid the first and obtain the second.

War between savage tribes, between Arabs, even today, [1] is a war of ambush by small groups of men of which each one, at the moment of surprise, chooses, not his adversary, but his victim, and is an assassin. Because the arms are similar on both sides, the only way of giving the advantage to one side is by surprise. A man surprised, needs an instant to collect his thoughts and defend himself; during this instant he is killed if he does not run away.

The surprised adversary does not defend himself, he tries to flee. Face to face or body to body combat with primitive arms, ax or dagger, so terrible among enemies without defensive arms, is very rare. It can take place only between enemies mutually surprised and without a chance of safety for any one except in victory. And still ... in case of mutual surprise, there is another chance of safety; that of falling back, of flight on the part of one or the other; and that chance is often seized. Here is an example, and if it does not concern savages at all, but soldiers of our days, the fact is none the less significant. It was observed by a man of warlike temperament who has related what he saw with his own eyes, although he was a forced spectator, held to the spot by a wound.

During the Crimean War, on a day of heavy fighting, two detachments of soldiers, A and B, coming around one of the mounds of earth that covered the country and meeting unexpectedly face to face, at ten paces, stopped thunderstruck. Then, forgetting their rifles, they threw stones and withdrew. Neither of the two groups had a decided leader to lead it to the front, and neither of the two dared to shoot first for fear that the other would at the same time bring his own arm to his shoulder. They were too near to hope to escape, or so they thought at least, although in reality, reciprocal firing, at such short ranges, is almost always too high. The man who would fire sees himself already killed by the return fire. He throws stones, and not with great force, to avoid using his rifle, to distract the enemy, to occupy the time, until flight offers him some chance of escaping at point-blank range.

This agreeable state of affairs did not last long, a minute perhaps. The appearance of a troop B on one flank determined the flight of A, and then the opposing group fired.

Surely, the affair is ridiculous and laughable.

Let us see, however. In a thick forest, a lion and a tiger meet face to face at a turn in the trail. They stop at once, rearing and ready to spring. They measure each other with their eyes, there is a rumbling in their throats. The claws move convulsively, the hair stands up. With tails lashing the ground, and necks stretched, ears flattened, lips turned up, they show their formidable fangs in that terrible threatening grimace of fear characteristic of felines.

Unseen, I shudder.

The situation is disagreeable for both: movement ahead means the death of a beast. Of which? Of both perhaps.

Slowly, quite slowly, one leg, bent for the leap, bending still, moves a few inches to the rear. Gently, quite gently, a fore paw follows the movement. After a stop, slowly, quite slowly, the other legs do the same, and both beasts, insensibly, little by little, and always facing, withdraw, up to the moment where their mutual withdrawal has created between them an interval greater than can be traversed in a bound. Lion and tiger turn their backs slowly and, without ceasing to observe, walk freely. They resume without haste their natural gaits, with that sovereign dignity characteristic of great seigneurs. I have ceased to shudder, but I do not laugh.

There is no more to laugh at in man in battle, because he has in his hands a weapon more terrible than the fangs and claws of lion or tiger, the rifle, which instantly, without possible defense, sends one from life into death. It is evident that no one close to his enemy is in a hurry to arm himself, to put into action a force which may kill him. He is not anxious to light the fuse that is to blow up the enemy, and himself at the same time.

Who has not observed like instances between dogs, between dog and cat, cat and cat?

In the Polish War of 1831, two Russian and two Polish regiments of cavalry charged each other. They went with the same dash to meet one another. When close enough to recognize faces, these cavalrymen slackened their gait and both turned their backs. The Russians and Poles, at this terrible moment, recognized each other as brothers, and rather than spill fraternal blood, they extricated themselves from a combat as if it were a crime. That is the version of an eyewitness and narrator, a Polish officer.

What do you think of cavalry troops so moved by brotherly love?

But let us resume:

When people become more numerous, and when the surprise of an entire population occupying a vast space is no longer possible, when a sort of public conscience has been cultivated within society, one is warned beforehand. War is formally declared. Surprise is no longer the whole of war, but it remains one of the means in war, the best means, even to-day. Man can no longer kill his enemy without defense. He has forewarned him. He must expect to find him standing and in numbers. He must fight; but he wishes to conquer with as little risk as possible. He employs the iron shod mace against the staff, arrows against the mace, the shield against arrows, the shield and cuirass against the shield alone, the long lance against the short lance, the tempered sword against the iron sword, the armed chariot against man on foot, and so on.

Man taxes his ingenuity to be able to kill without running the risk of being killed. His bravery is born of his strength and it is not absolute. Before a stronger he flees without shame. The instinct of self-preservation is so powerful that he does not feel disgraced in obeying it, although, thanks to the defensive power of arms and armor he can fight at close quarters. Can you expect him to act in any other way? Man must test himself before acknowledging a stronger. But once the stronger is recognized, no one will face him.

Individual strength and valor were supreme in primitive combats, so much so that when its heroes were killed, the nation was conquered. As a result of a mutual and tacit understanding, combatants often stopped fighting to watch with awe and anxiety two champions struggling. Whole peoples often placed their fate in the hands of the champions who took up the task and who alone fought. This was perfectly natural. They counted their champion a superman, and no man can stand against the superman.

But intelligence rebels against the dominance of force. No one can stand against an Achilles, but no Achilles can withstand ten enemies who, uniting their efforts, act in concert. This is the reason for tactics, which prescribe beforehand proper means of organization and action to give unanimity to effort, and for discipline which insures united efforts in spite of the innate weakness of combatants.

In the beginning man battled against man, each one for himself, like a beast that hunts to kill, yet flees from that which would kill him. But now prescriptions of discipline and tactics insure unity between leader and soldier, between the men themselves. Besides the intellectual progress, is there a moral progress? To secure unity in combat, to make tactical dispositions in order to render it practically possible, we must be able to count on the devotion of all. This elevates all combatants to the level of the champions of primitive combat. Esprit appears, flight is a disgrace, for one is no longer alone in combat. There is a legion, and he who gives way quits his commanders and his companions. In all respects the combatant is worth more.

So reason shows us the strength of wisely united effort; discipline makes it possible.

Will the result be terrible fights, conflicts of extermination? No! Collective man, a disciplined body of troops formed in tactical battle order, is invincible against an undisciplined body of troops. But against a similarly disciplined body, he becomes again primitive man. He flees before a greater force of destruction when he recognizes it or when he foresees it. Nothing is changed in the heart of man. Discipline keeps enemies face to face a little longer, but cannot supplant the instinct of self-preservation and the sense of fear that goes with it.

Fear!...

There are officers and soldiers who do not know it, but they are people of rare grit. The mass shudders; because you cannot suppress the flesh. This trembling must be taken into account in all organization, discipline, arrangements, movements, maneuvers, mode of action. All these are affected by the human weakness of the soldier which causes him to magnify the strength of the enemy.

This faltering is studied in ancient combat. It is seen that of nations apt in war, the strongest have been those who, not only best have understood the general conduct of war, but who have taken human weakness into greatest account and taken the best guarantees against it. It is notable that the most warlike peoples are not always those in which military institutions and combat methods are the best or the most rational.

And indeed, in warlike nations there is a good dose of vanity. They only take into account courage in their tactics. One might say that they do not desire to acknowledge weakness.

The Gaul, a fool in war, used barbarian tactics. After the first surprise, he was always beaten by the Greeks and Romans.

The Greek, a warrior, but also a politician, had tactics far superior to those of the Gauls and the Asiatics.

The Roman, a politician above all, with whom war was only a means, wanted perfect means. He had no illusions. He took into account human weakness and he discovered the legion.

But this is merely affirming what should be demonstrated.

CHAPTER II

KNOWLEDGE OF MAN MADE ROMAN TACTICS.

THE SUCCESSES OF HANNIBAL, THOSE OF CAESAR

Greek tactics developed the phalanx; Roman tactics, the legion; the tactics of the barbarians employed the square phalanx, wedge or lozenge.

The mechanism of these various formations is explained in all elementary books. Polybius enters into a mechanical discussion when he contrasts the phalanx and the legion. (Book 18.)

The Greeks were, in intellectual civilization, superior to the Romans, consequently their tactics ought to have been far more rational. But such was not the case. Greek tactics proceeded from mathematical reasoning; Roman tactics from a profound knowledge of man's heart. Naturally the Greeks did not neglect morale nor the Romans mechanics, [2] but their primary, considerations were diverse.

What formation obtained the maximum effort from the Greek army?

What methods caused the soldiers of a Roman army to fight most effectively?

The first question admits of discussion. The Roman solved the second.

The Roman was not essentially brave. He did not produce any warrior of the type of Alexander. It is acknowledged that the valorous impetuosity of the barbarians, Gauls, Cimbri, Teutons, made him tremble. But to the glorious courage of the Greeks, to the natural bravery of the Gauls he opposed a strict sense of duty, secured by a terrible discipline in the masses. It was inspired in the officers by a sentiment of the strongest patriotism.

The discipline of the Greeks was secured by exercises and rewards; the discipline of the Romans was secured also by the fear of death. They put to death with the club; they decimated their cowardly or traitorous units.

In order to conquer enemies that terrified his men, a Roman general heightened their morale, not by enthusiasm but by anger. He made the life of his soldiers miserable by excessive work and privations. He stretched the force of discipline to the point where, at a critical instant, it must break or expend itself on the enemy. Under similar circumstances, a Greek general caused Tyrtaeus to sing. [3] It would have been curious to see two such forces opposed.

But discipline alone does not constitute superior tactics. Man in battle, I repeat, is a being in whom the instinct of self-preservation dominates, at certain moments, all other sentiments. Discipline has for its aim the domination of that instinct by a greater terror. But it cannot dominate it completely. I do not deny the glorious examples where discipline and devotion have elevated man above himself. But if these examples are glorious, it is because they are rare; if they are admired, it is because they are considered exceptions, and the exception proves the rule.

The determination of that instant where man loses his reasoning power and becomes instinctive is the crowning achievement in the science of combat. In general, here was the strength of the Roman tactics. In particular cases such successful determination makes Hannibals and Caesars.

Combat took place between masses in more or less deep formation commanded and supervised by leaders with a definite mission. The combat between masses was a series of individual conflicts, juxtaposed, with the front rank man alone fighting. If he fell, if he was wounded or worn out, he was replaced by the man of the second rank who had watched and guarded his flanks. This procedure continued up to the last rank. Man is always physically and morally fatigued in a hand-to-hand tournament where he employs all his energy.

These contests generally lasted but a short time. With like morale, the least fatigued always won.

During this engagement of the first two ranks, the one fighting, the other watching close at hand, the men of the rear ranks waited inactive at two paces distance for their turn in the combat, which would come only when their predecessors were killed, wounded or exhausted. They were impressed by the violent fluctuations of the struggle of the first rank. They heard the clashes of the blows and distinguished, perhaps, those that sank into the flesh. They saw the wounded, the exhausted crawl through the intervals to go to the rear. Passive spectators of danger, they were forced to await its terrible approach. These men were subjected to the poignant emotions of combat without being supported by the animation of the struggle. They were thus placed under the moral pressure of the greatest of anxieties. Often they could not stand it until their turn came; they gave way.

The best tactics, the best dispositions were those that made easiest a succession of efforts by assuring the relief by ranks of units in action, actually engaging only the necessary units and keeping the rest as a support or reserve outside of the immediate sphere of moral tension. The superiority of the Romans lay in such tactics and in the terrible discipline which prepared and assured the execution. By their resistance against fatigue which rude and continual tasks gave them and by the renewal of combatants in combat, they secured greater continuity of effort than any others. [4]

The Gauls did not reason. Seeing only the inflexible line, they bound themselves together, thus rendering relief impracticable. They believed, as did the Greeks, in the power of the mass and impulse of deep files, and did not understand that deep files were powerless to push the first ranks forward as they recoiled in the face of death. It is a strange error to believe that the last ranks will go to meet that which made the first ones fall back. On the contrary, the contagion of recoil is so strong that the stopping of the head means the falling back of the rear!

The Greeks, also, certainly had reserves and supports in the second half of their dense ranks. But the idea of mass dominated. They placed these supports and reserves too near, forgetting the essential, man.

The Romans believed in the power of mass, but from the moral point of view only. They did not multiply the files in order to add to the mass, but to give to the combatants the confidence of being aided and relieved. The number of ranks was calculated according to the moral pressure that the last ranks could sustain.

There is a point beyond which man cannot bear the anxiety of combat in the front lines without being engaged. The Romans did not so increase the number of ranks as to bring about this condition. The Greeks did not observe and calculate so well. They sometimes brought the number of files up to thirty-two and their last files, which in their minds, were doubtless their reserves, found themselves forcibly dragged into the material disorder of the first ones.

In the order by maniples in the Roman legion, the best soldiers, those whose courage had been proved by experience in battle, waited stoically, kept in the second and third lines. They were far enough away not to suffer wounds and not to be drawn in by the front line retiring into their intervals. Yet they were near enough to give support when necessary or to finish the job by advancing.

When the three separate and successive maniples of the first cohort were united in order to form the united battle cohort of Marius and of Caesar, the same brain placed the most reliable men in the last lines, i.e., the oldest. The youngest, the most impetuous, were in the first lines. The legion was not increased simply to make numbers or mass. Each had his turn in action, each man in his maniple, each maniple in its cohort, and, when the unit became a cohort, each cohort in the order of battle.

We have seen that the Roman theory dictated a depth of ranks to furnish successive lines of combatants. The genius of the general modified these established formations. If the men were inured to war, well-trained, reliable, tenacious, quick to relieve their file leaders, full of confidence in their general and their own comrades, the general diminished the depth of the files, did away with the lines even, in order to increase the number of immediate combatants by increasing the front. His men having a moral, and sometimes also a physical endurance superior to that of the adversary, the general knew that the last ranks of the latter would not, under pressure, hold sufficiently to relieve the first lines nor to forbid the relief of his own. Hannibal had a part of his infantry, the Africans, armed and drilled in the Roman way; his Spanish infantrymen had the long wind of the Spaniards of to-day; his Gallic soldiers, tried out by hardship, were in the same way fit for long efforts. Hannibal, strong with the confidence with which he inspired his people, drew up a line less deep by half than the Roman army and at Cannae hemmed in an army which had twice his number and exterminated it. Caesar at Pharsalus, for similar reasons, did not hesitate to decrease his depth. He faced double his strength in the army of Pompey, a Roman army like his own, and crushed it.

We have mentioned Cannae and Pharsalus, we shall study in them the mechanism and the morale of ancient combat, two things which cannot be separated. We cannot find better examples of battle more clearly and more impartially exhibited. This is due in one case to the clear presentation of Polybius, who obtained his information from the fugitives from Cannae, possibly even from some of the conquerors; in the other it is due to the impassive clearness of Caesar in describing the art of war.

CHAPTER III

ANALYSIS OF THE BATTLE OF CANNAE

Recital of Polybius:
"Varro placed the cavalry on the right wing, and rested it on the river; the infantry was deployed near it and on the same line, the maniples drawn close to each other, with smaller intervals than usual, and the maniples presenting more depth than front.

"The cavalry of the allies, on the left wing, completed the line, in front of which were posted the light troops. There were in that army, including the allies, eighty thousand foot and a little more than six thousand horse.

"Meanwhile Hannibal had his slingers and light troops cross the Aufidus and posted them in front of his army. The rest crossed the river at two places. He placed the Iberian and Gallic cavalry on the left wing, next the river and facing the Roman cavalry. He placed on the same line, one half of the African infantry heavily armed, the Iberian and Gallic infantry, the other half of the African infantry, and finally the Numidian cavalry which formed the right wing.

"After he had thus arrayed all his troops upon a single line, he marched to meet the enemy with the Iberian and Gallic infantry moving independently of the main body. As it was joined in a straight line with the rest, on separating, it was formed like the convex face of a crescent. This formation reduced its depth in the center. The intention of the general was to commence the battle with the Iberians and Gauls, and have them supported by the Africans.

"The latter infantry was armed like the Roman infantry, having been equipped by Hannibal with arms that had been taken from the Romans in preceding battle. Both Iberians and Gauls had shields; but their swords were quite different. The sword of the former was as fit for thrusting as for cutting while that of the Gauls only cut with the edge, and at a limited distance. These troops were drawn up as follows: the Iberians were in two bodies of troops on the wings, near the Africans; the Gauls in the center. The Gauls were nude; the Iberians in linen shirts of purple color, which to the Romans was an extraordinary and frightening spectacle. The Carthaginian army consisted of ten thousand horse and little more than forty thousand foot.

"Aemilius commanded the right of the Romans, Varro the left; the two consuls of the past year, Servilius and Attilius, were in the center. On the Carthaginian side, Hasdrubal had the left under his orders, Hanno the right, and Hannibal, who had his brother Mago with him, reserved for himself the command of the center. The two armies did not suffer from the glare of the sun when it rose, the one being faced to the South, as I remarked, and the other to the North.

"Action commenced with the light troops, which were in front of both armies. The first engagement gave advantage to neither the one nor the other. Just as soon as the Iberian and Gallic cavalry on the left approached, the conflict became hot. The Romans fought with fury and rather more like barbarians than

Romans. This falling back and then returning to the charge was not according to their tactics. Scarcely did they become engaged when they leaped from their horses and each seized his adversary. In the meanwhile the Carthaginians gained the upper hand. The greater number of the Romans remained on the ground after having fought with the greatest valor. The others were pursued along the river and cut to pieces without being able to obtain quarter.

"The heavily armed infantry immediately took the place of the light troops and became engaged. The Iberians and Gauls held firm at first and sustained the shock with vigor; but they soon gave way to the weight of the legions, and, opening the crescent, turned their backs and retreated. The Romans followed them with impetuosity, and broke the Gallic line much more easily because the wings crowded toward the center where the thick of the fighting was. The whole line did not fight at the same time. The action commenced in the center because the Gauls, being drawn up in the form of a crescent, left the wings far behind them, and presented the convex face of the crescent to the Romans. The latter then followed the Gauls and Iberians closely, and crowded towards the center, to the place where the enemy gave way, pushing ahead so forcibly that on both flanks they engaged the heavily armed Africans. The Africans on the right, in swinging about from right to left, found themselves all along the enemy's flank, as well as those on the left which made the swing from left to right. The very circumstances of the action showed them what they had to do. This was what Hannibal had foreseen; that the Romans pursuing the Gauls must be enveloped by the Africans. The Romans then, no longer able to keep their formation [5] were forced to defend themselves man to man and in small groups against those who attacked them on front and flank.[6]

"Aemilius had escaped the carnage on the right wing at the commencement of the battle. Wishing, according to the orders he had given, to be everywhere, and seeing that it was the legionary infantry that would decide the fate of the battle, he pushed his horse through the fray, warded off or killed every one who opposed him, and sought at the same time to reanimate the ardor of the Roman soldiers. Hannibal, who during the entire battle remained in the conflict, did the same in his army.

"The Numidian cavalry on the right wing, without doing or suffering much, was useful on that occasion by its manner of fighting; for, pouncing upon the enemy on all sides, they gave him enough to do so that he might not have time to think of helping his own people. Indeed, when the left wing, where Hasdrubal commanded, had routed almost all the cavalry of the Roman right wing, and a junction had been effected with the Numidians, the auxiliary cavalry did not wait to be attacked but gave way.

"Hasdrubal is said to have done something which proved his prudence and his ability, and which contributed to the success of the battle. As the Numidians were in great number, and as these troops were never more useful than when one was in flight before them, he gave them the fugitives to pursue, and led the Iberian and Gallic cavalry in a charge to aid the African infantry. He pounced on the Romans from the rear, and having bodies of cavalry charge into the mêlée at several places, he gave new strength to the Africans and made the arms drop from the hands of the adversaries. It was then that L. Aemilius, a citizen who during his whole life, as in this last conflict, had nobly fulfilled his duties to his country, finally succumbed, covered with mortal wounds.

"The Romans continued fighting, giving battle to those who were surrounding them. They resisted to the last. But as their numbers diminished more and more, they were finally forced into a smaller circle, and all put to the sword. Attilius and Servilius, two persons of great probity, who had distinguished themselves in the combat as true Romans, were also killed on that occasion.

"While this carnage was taking place in the center, the Numidians pursued the fugitives of the left wing. Most of them were cut down, others were thrown under their horses; some of them escaped to Venusia. Among these was Varro, the Roman general, that abominable man whose administration cost his country so dearly. Thus ended the battle of Cannae, a battle where prodigies of valor were seen on both sides.

"Of the six thousand horse of which the Roman cavalry was composed, only seventy Romans reached Venusia with Varro, and, of the auxiliary cavalry, only three hundred men found shelter in various towns. Ten thousand foot were taken prisoners, but they were not in the battle. [7] Of troops in battle only about three thousand saved themselves in the nearby town; the balance, numbering about twenty thousand, died on the field of honor." [8]

Hannibal lost in that action in the neighborhood of four thousand Gauls, fifteen hundred Iberians and Africans and two hundred horses.

Let us analyze:

The light infantry troops were scattered in front of the armies and skirmished without result. The real combat commenced with the attack on the legitimate cavalry of the Roman left wing by the cavalry of Hannibal.

There, says Polybius, the fight grew thickest, the Romans fought with fury and much more like barbarians than like Romans; because this falling back, then returning to the charge was not according to their tactics; scarcely did they become engaged when they leaped from their horses and each seized his adversary, etc., etc.

This means that the Roman cavalry did not habitually fight hand to hand like the infantry. It threw itself in a gallop on the enemy cavalry. When within javelin range, if the enemy's cavalry had not turned in the opposite direction on seeing the Roman cavalry coming, the latter prudently slackened its gait, threw some javelins, and, making an about by platoons, took to the rear for the purpose of repeating the charge. The hostile cavalry did the same, and such an operation might be renewed several times, until one of the two, persuaded that his enemy was going to attack him with a dash, turned in flight and was pursued to the limit.

That day, the fight becoming hot, they became really engaged; the two cavalry bodies closed and man fought man. The fight was forced, however; as there was no giving way on one side or the other, it was necessary actually to attack. There was no space for skirmishing. Closed in by the Aufidus and the legions, the Roman cavalry could not operate (Livy). The Iberian and Gallic cavalry, likewise shut in and double the Roman cavalry, was forced into two lines; it could still less maneuver. This limited front served the Romans, inferior in number, who could thus be attacked only in front, that is by an equal number. It rendered, as we have said, contact inevitable. These two cavalry bodies placed chest to chest had to fight close, had to grapple man to man, and for riders mounted on simple saddle cloths and without stirrup, embarrassed with a shield, a lance, a saber or a sword, to grapple man to man is to grapple together, fall together and fight on foot. That is what happened, as the account of Titus Livius explains it in completing that of Polybius. The same thing happened every time that two ancient cavalry organizations really had to fight, as the battle of the Tecinus showed. This mode of action was all to the advantage of the Romans, who were well-armed and well-trained therein. Note the battle of Tecinus. The Roman light infantry was cut to pieces, but the elite of the Roman cavalry, although surprised and surrounded, fought a-foot and on horse back, inflicted more casualties on the cavalry of Hannibal than they suffered, and brought back from the field their wounded general. The Romans besides were well led by Consul Aemilius, a man of head and heart, who, instead of fleeing when his cavalry was defeated, went himself to die in the ranks of the infantry.

Meanwhile we see thirty to thirty-four hundred Roman cavalrymen nearly exterminated by six to seven thousand Gauls and Iberians who did not lose even two hundred men. Hannibal's entire cavalry lost but two hundred men on that day.

How can that be explained?

Because most of them died without dreaming of selling their lives and because they took to flight during the fight of the first line and were struck with impunity from behind. The words of Polybius: "Most of them remained on the spot after having defended themselves with the utmost valor," were consecrated words before Polybius. The conquered always console themselves with their bravery and conquerors never contradict. Unfortunately, the figures are there. The facts of the battle are found in the account, which sounds no note of desperation. The Gallic and Roman cavalry had each already made a brave effort by attacking each other from the front. This effort was followed by the terrible anxiety of close combat. The Roman cavalrymen, who from behind the combatants on foot were able to see the second Gallic line on horse back, gave ground. Fear very quickly made the disengaged ranks take to their horses, wheel about like a flock of sheep in a stampede, and abandon their comrades and themselves to the mercy of the conquerors.

Yet, these horsemen were brave men, the elite of the army, noble knights, guards of the consuls, volunteers of noble families.

The Roman cavalry defeated, Hasdrubal passed his Gallic and Iberian troopers behind Hannibal's army, to attack the allied cavalry till then engaged by the Numidians. [9] The cavalry of the allies did not await the enemy. It turned its back immediately; pursued to the utmost by the Numidians who were numerous (three thousand), and excellent in pursuit, it was reduced to some three hundred men, without a struggle.

After the skirmishing of the light infantry troops, the foot-soldiers of the line met. Polybius has explained to us how the Roman infantry let itself be enclosed by the two wings of the Carthaginian army and taken in rear by Hasdrubal's cavalry. It is also probable that the Gauls and Iberians, repulsed in the first part of the action and forced to turn their backs, returned, aided by a portion of the light infantry, to the charge upon the apex of the wedge formed by the Romans and completed their encirclement.

But we know, as will be seen further on in examples taken from Caesar, that the ancient cavalryman was powerless against formed infantry, even against the isolated infantryman possessing coolness. The Iberian and Gallic cavalry ought to have found behind the Roman army the reliable triarians penned in, armed, with pikes. [10] It might have held them in check, forced them to give battle, but done them little or no harm as long as the ranks were preserved.

We know that of Hannibal's infantry only twelve thousand at the most were equipped with Roman weapons. We know that his Gallic and Iberian infantry, protected by plain shields, had to fall back, turn, and probably lost in this part of the action very nearly the four thousand men, which the battle cost them.

Let us deduct the ten thousand men that had gone to the attack of Hannibal's camp and the five thousand which the latter must have left there. There remain:

A mass of seventy thousand men surrounded and slaughtered by twenty-eight thousand foot soldiers, or, counting Hasdrubal's cavalry, by thirty-six thousand men, by half their number.

It may be asked how seventy thousand men could have let themselves be slaughtered, without defense, by thirty-six thousand men less well-armed, when each combatant had but one man before him. For in close combat, and especially in so large an envelopment, the number of combatants immediately engaged was the same on each side. Then there were neither guns nor rifles able to pierce the mass by a converging fire and destroy it by the superiority of this fire over diverging fire. Arrows were exhausted in the first period of the action. It seems that, by their mass, the Romans must have presented an insurmountable resistance, and that while permitting the enemy to wear himself out against it, that mass had only to defend itself in order to repel assailants.

But it was wiped out.

In pursuit of the Gauls and Iberians, who certainly were not able, even with like morale, to stand against the superior arms of the legionaries, the center drove all vigorously before it. The wings, in order to support it and not to lose the intervals, followed its movement by a forward oblique march and formed the sides of the salient. The entire Roman army, in wedge order, marched to victory. Suddenly the wings were attacked by the African battalions; the Gauls, the Iberians, [11] who had been in retreat, returned to the fight. The horsemen of Hasdrubal, in the rear, attacked the reserves. [12] Everywhere there was combat, unexpected, unforeseen. At the moment when they believed themselves conquerors, everywhere, in front, to the right, to the left, in the rear, the Roman soldiers heard the furious clamor of combat. [13]

The physical pressure was unimportant. The ranks that they were fighting had not half their own depth. The moral pressure was enormous. Uneasiness, then terror, took hold of them; the first ranks, fatigued or wounded, wanted to retreat; but the last ranks, frightened, withdrew, gave way and whirled into the interior of the wedge. Demoralized and not feeling themselves supported, the ranks engaged followed them, and the routed mass let itself be slaughtered. The weapons fell from their hands, says Polybius.

The analysis of Cannae is ended. Before passing to the recital of Pharsalus, we cannot resist the temptation, though the matter be a little foreign to the subject, to say a few words about the battles of Hannibal.

These battles have a particular character of stubbornness explained by the necessity for overcoming the Roman tenacity. It may be said that to Hannibal victory was not sufficient. He must destroy. Consequently he always tried to cut off all retreat for the enemy. He knew that with Rome, destruction was the only way of finishing the struggle.

He did not believe in the courage of despair in the masses; he believed in terror and he knew the value of surprise in inspiring it.

But it was not the losses of the Romans that was the most surprising thing in these engagements. It was the losses of Hannibal. Who, before Hannibal or after him, has lost as many as the Romans and yet been conqueror? To keep troops in action, until victory comes, with such losses, requires a most powerful hand.

He inspired his people with absolute confidence. Almost always his center, where he put his Gauls, his food for powder, was broken. But that did not seem to disquiet or trouble either him or his men.

It is true that his center was pierced by the Romans who were escaping the pressure of the two Carthaginian wings, that they were in disorder because they had fought and pushed back the Gauls, whom Hannibal knew how to make fight with singular tenacity. They probably felt as though they had escaped from a press, and, happy to be out of it, they thought only of getting further away from the battle and by no means of returning to the flanks or the rear of the enemy. In addition, although nothing is said about it, Hannibal had doubtless taken precautions against their ever returning to the conflict.

All that is probably true. The confidence of the Gallic troops, so broken through, is none the less surprising.

Hannibal, in order to inspire his people with such confidence, had to explain to them before the combat his plan of action, in such a way that treachery could not injure him. He must have warned his troops that the center would be pierced, but that he was not worried about it, because it was a foreseen and prepared affair. His troops, indeed, did not seem to be worried about it.

Let us leave aside his conception of campaigns, his greatest glory in the eyes of all. Hannibal was the greatest general of antiquity by reason of his admirable comprehension of the morale of combat, of the morale of the soldier whether his own or the enemy's. He shows his greatness in this respect in all the different incidents of war, of campaign, of action. His men were not better than the Roman soldiers. They were not as well armed, one-half less in number. Yet he was always the conqueror. He understood the value of morale. He had the absolute confidence of his people. In addition he had the art, in commanding an army, of always securing the advantage of morale.

In Italy he had, it is true, cavalry superior to that of the Romans. But the Romans had a much superior infantry. Had conditions been reversed, he would have changed his methods. The instruments of battle are valuable only if one knows how to use them, and Pompey, we shall see, was beaten at Pharsalus precisely because he had a cavalry superior to that of Caesar.

If Hannibal was vanquished at Zuma, it was because genius cannot accomplish the impossible. Zuma proved again the perfect knowledge of men that Hannibal possessed and his influence over the troops. His third line, the only one where he really had reliable soldiers, was the only one that fought. Beset on all sides, it slew two thousand Romans before it was conquered.

We shall see later what a high state of morale, what desperate fighting, this meant.

CHAPTER IV

ANALYSIS OF THE BATTLE OF PHARSALUS, AND SOME CHARACTERISTIC EXAMPLES

Here is Caesar's account of the battle of Pharsalus.

"As Caesar approached Pompey's camp, he noted that Pompey's army was placed in the following order:

"On the left wing were the 2nd and 3rd Legions which Caesar had sent to Pompey at the commencement of the operation, pursuant to a decree of the Senate, and which Pompey had kept. Scipio occupied the center with the legions from Syria. The legion from Cilicia was placed on the right wing together with the Spanish cohorts of Afranius. Pompey regarded the troops already mentioned as the most reliable of his army. Between them, that is, between the center and the wings, he had distributed the remainder, consisting of one hundred and ten complete cohorts in line. These were made up of forty-five thousand men, two thousand of whom were veterans, previously rewarded for their services, who had come to join him. He had scattered them throughout the whole line of battle. Seven cohorts had been left to guard his camp and the neighboring forts. His right wing rested on a stream with inaccessible banks; and, for that reason, he had placed all his seven thousand cavalry, [14] his archers and his slingers (forty-two hundred men) on the left wing.

"Caesar, keeping his battle order, [15] had placed the 10th Legion on the right wing, and on the left, the 9th, which was much weakened by the combats of Dyrrachium. To the latter he added the 8th in order to form something like a full legion from the two, and ordered them to support one another. He had eighty very completely organized cohorts in line, approximately twenty-two thousand men. Two cohorts had been left to guard the camp. Caesar had entrusted the command of the left wing to Anthony, that of the right to P. Sylla, and of the center to C. Domitius. He placed himself in front of Pompey. But when he saw the disposition of the opposing army, he feared that his right wing was going to be enveloped by Pompey's numerous cavalry. He therefore withdrew immediately from his third line a cohort from each legion (six cohorts), in order to form a fourth line, placed it to receive Pompey's cavalry and showed it what it had to do. Then he explained fully to these cohorts that the success of the day depended on their valor. At the same time he ordered the entire army, and in particular the third line, not to move without his command, reserving to himself authority to give the signal by means of the standard when he thought it opportune.

"Caesar then went through his lines to exhort his men to do well, and seeing them full of ardor, had the signal given.

"Between the two armies there was only enough space to give each the necessary distance for the charge. But Pompey had given his men orders to await the charge without stirring, and to let Caesar's army break its ranks upon them. He did this, they say, on the advice of C. Triarius, as a method of meeting the force of the first dash of Caesar's men. He hoped that their battle order would be broken up and his own soldiers, well disposed in ranks, would have to fight with sword in hand only men in disorder. He thought that this formation would best protect his troops from the force of the fall of heavy javelins. At the same time he hoped that Caesar's soldiers charging at the run would be out of breath and overcome with fatigue at the moment of contact. Pompey's immobility was an error because there is in every one an animation, a natural ardor that is instilled by the onset to the combat. Generals ought not to check but to encourage this ardor. It was for this reason that, in olden times, troops charged with loud shouts, all trumpets sounding, in order to frighten the enemy and encourage themselves.

"In the meanwhile, our soldiers, at the given signal advanced with javelins in hand; but having noticed that Pompey's soldiers were not running towards them, and taught by experience and trained by previous battles, they slowed down and stopped in the midst of their run, in order not to arrive out of breath and worn out. Some moments after, having taken up their run again, they launched their javelins, and immediately afterwards, according to Caesar's order drew their swords. The Pompeians conducted themselves perfectly. They received the darts courageously; they did not stir before the dash of the legions; they preserved their lines, and, having dispatched their javelins, drew their swords.

"At the same time Pompey's entire cavalry dashed from the left wing, as had been ordered, and the mass of his archers ran from all parts of the line. Our cavalry did not await the charge, but fell back a little. Pompey's cavalry became more pressing, and commenced to reform its squadrons and turn our exposed flank. As soon as Caesar saw this intention, he gave the signal to the fourth line of six cohorts. This line started directly and, standards low, they charged the Pompeian cavalry with such vigor and resolution that not a single man stood his ground. All wheeled about and not only withdrew in full flight, but gained the highest mountains as fast as they could. They left the archers and slingers without their defense and protection. These were all killed. At the same time the cohorts moved to the rear of Pompey's left wing, which was still fighting and resisting, and attacked it in rear.

"Meanwhile, Caesar had advanced his third line, which up to this moment had been kept quietly at its post. These fresh troops relieved those that were fatigued. Pompey's men, taken in rear, could no longer hold out and all took to flight.

"Caesar was not in error when he put these cohorts in a fourth line, particularly charged with meeting the cavalry, and urged them to do well, since their effort would bring victory. They repulsed the cavalry. They cut to pieces the slingers and archers. They turned Pompey's left wing, and this decided the day.

"When Pompey saw his cavalry repulsed and that portion of the army upon which he had counted the most seized with terror, he had little confidence in the rest. He quit the battle and galloped to his camp, where, addressing his centurions who were guarding the praetorian gate, he told them in a loud voice heard by the soldiers: 'Guard well the camp and defend it vigorously in case of attack; as for myself, I am going to make the tour of the other gates and assure their defense.'

"That said, he retired to the praetorium, despairing of success and awaiting events.

"After having forced the enemy to flee to his entrenchments Caesar, persuaded that he ought not to give the slightest respite to a terrorized enemy, incited his soldiers to profit by their advantage and attack the camp. Although overcome by the heat, for the struggle was prolonged into the middle of the day, they did not object to greater fatigue and obeyed. The camp was at first well defended by the cohorts on watch and especially by the Thracians and barbarians. The men who had fled from the battle, full of fright and overcome with fatigue, had nearly all thrown their arms and colors away and thought rather more of saving themselves than of defending the camp. Even those who defended the entrenchments were unable long to resist the shower of arrows. Covered with wounds, they abandoned the place, and led by their centurions and tribunes, they took refuge as quickly as they could in the high mountains near the camp.

"Caesar lost in this battle but two hundred soldiers, but nearly thirty of the bravest centurions were killed therein. Of Pompey's army fifteen thousand perished, and more than twenty-four thousand took refuge in the mountains. As Caesar had invested the mountains with entrenchments, they surrendered the following day."

Such is Caesar's account. His action is so clearly shown that there is scarcely any need of comment.

Initially Caesar's formation was in three lines. This was the usual battle order in the Roman armies, without being absolute, however, since Marius fought with two only. But, as we have said, according to the occasion, the genius of the chief decided the battle formation. There is no reason to suppose that Pompey's army was in a different order of battle.

To face that army, twice as large as his, Caesar, if he had had to preserve the disposition of cohorts in ten ranks, would have been able to form but one complete line, the first, and a second, half as numerous, as a reserve. But he knew the bravery of his troops, and he knew the apparent force of deep ranks to be a delusion. He did not hesitate to diminish his depth in order to keep the formation and morale of three-fifths of his troops intact, until the moment of their engagement. In order to be even more sure of the third line of his reserve, and in order to make sure that it would not be carried away by its enthusiasm for action, he paid it most particular attention. Perhaps, the text is doubtful, he kept it at double the usual distance in rear of the fighting lines.

Then, to guard against a turning movement by Pompey's seven thousand cavalry and forty-two hundred slingers and archers, a movement in which Pompey placed the hopes of victory, Caesar posted six cohorts that represented scarcely two thousand men. He had perfect confidence that these two thousand men would make Pompey's cavalry wheel about, and that his one thousand horsemen would then press the action so energetically that Pompey's cavalry would not even think of rallying. It happened so; and the forty-two hundred archers and slingers were slaughtered like sheep by these cohorts, aided, without doubt, by four-hundred foot [16] young and agile, whom Caesar mixed with his thousand horsemen and who remained at this task, leaving the horsemen, whom they had relieved, to pursue the terror-stricken fugitives.

Thus were seven thousand horsemen swept away and forty-two hundred infantrymen slaughtered without a struggle, all demoralized simply by a vigorous demonstration.

The order to await the charge, given by Pompey to his infantry, was judged too severely by Caesar. Caesar certainly was right as a general rule; the enthusiasm of the troops must not be dampened, and the initiative of the attack indeed gives to the assailant a certain moral influence. But with trusted soldiers, duly trained, one can try a stratagem, and the men of Pompey had proven their dependability by awaiting on the spot, without stirring, a vigorous enemy in good order, when they counted on meeting him in disorder and out of breath. Though it may not have led to success, the advice of Triarius was not bad. Even the conduct of Caesar's men proves this. This battle shows the confidence of the soldier in the material rank in ancient combat, as assuring support and mutual assistance.

Notwithstanding the fact the Caesar's soldiers had the initiative in the attack, the first encounter decided nothing. It was a combat on the spot, a struggle of several hours. Forty-five thousand good troops lost scarcely two hundred men in this struggle for, with like arms, courage and ability, Pompey's infantry ought not to have lost in hand-to-hand fighting more than that of Caesar's. These same forty-five thousand men gave way, and, merely between the battle field and their camp, twelve thousand were slaughtered.

Pompey's men had twice the depth of Caesar's ranks, whose attack did not make them fall back a step. On the other hand their mass was unable to repel him, and he was fought on the spot. Pompey had announced to them, says Caesar, that the enemy's army would be turned by his cavalry, and suddenly, when they were fighting bravely, step by step, they heard behind them the shouts of attack by the six cohorts of Caesar, two thousand men.

Does it seem an easy matter for such a force to ward off this menace? No. The wing taken in rear in this way loses ground; more and more the contagion of fear spreads to the rest. Terror is so great that they do not think of re-forming in their camp, which is defended for a moment only by the cohorts on guard. Just as at Cannae, their arms drop from their hands. But for the good conduct of the camp guards which permitted the fugitives to gain the mountains, the twenty-four thousand prisoners of the next day might have been corpses that very day.

Cannae and Pharsalus, are sufficient to illustrate ancient combat. Let us, however, add some other characteristic examples, which we shall select briefly and in chronological order. They will complete our data. [17]

Livy relates that in an action against some of the peoples in the neighborhood of Rome, I do not recall now which, the Romans did not dare to pursue for fear of breaking their ranks.

In a fight against the Hernici, he cites the Roman horsemen, who had not been able to do anything on horseback to break up the enemy, asking the consul for permission to dismount and fight on foot. This is true not only of Roman cavalrymen, for later on we shall see the best riders, the Gauls, the Germans, the Parthanians even, dismounting in order really to fight.

The Volsci, the Latini, the Hernici, etc., combined to fight the Romans; and as the action nears its end, Livy relates: "Finally, the first ranks having fallen, and carnage being all about them, they threw away their arms and started to scatter. The cavalry then dashed forward, with orders not to kill the isolated ones, but to harass the mass with their arrows, annoy it, to delay it, to prevent dispersion in order to permit the infantry to come up and kill."

In Hamilcar's engagement against the mercenaries in revolt, who up to then had always beaten the Carthaginians, the mercenaries endeavored to envelop him. Hamilcar surprised them by a new maneuver and defeated them. He marched in three lines: elephants, cavalry and light infantry, then heavily armed phalanxes. At the approach of the mercenaries who were marching vigorously towards him the two lines formed by the elephants, the cavalry and light infantry, turned about and moved quickly to place themselves on the flanks of the third line. The third line thus exposed met a foe which had thought only of pursuit, and which the surprise put to flight. It thus abandoned itself to the action of the elephants, horses and the light infantry who massacred the fugitives.

Hamilcar killed six thousand men, captured two thousand and lost practically nobody. It was a question as to whether he had lost a single man, since there had been no combat.

In the battle of Lake Trasimenus, the Carthaginians lost fifteen hundred men, nearly all Gauls; the Romans fifteen thousand and fifteen thousand prisoners. The battle raged for three hours.

At Zama, Hannibal had twenty thousand killed, twenty thousand prisoners; the Romans two thousand killed. This was a serious struggle in which Hannibal's third line alone fought. It gave way only under the attack on its rear and flank by the cavalry.

In the battle of Cynoscephalae, between Philip and Flaminius, Philip pressed Flaminius with his phalanx thirty-two deep. Twenty maniples took the phalanx from behind. The battle was lost by Philip. The Romans had seven hundred killed; the Macedonians eighty thousand, and five thousand prisoners.

At Pydna, Aemilius Paulus against Perseus, the phalanx marched without being stopped. But gaps occurred from the resistance that it encountered. Hundreds penetrated into the gaps in the phalanx and killed the men embarrassed with their long pikes. They were effective only when united, abreast, and at shaft's length. There was frightful disorder and butchery; twenty thousand killed, five thousand captured out of forty-four thousand engaged! The historian does not deem it worth while to speak of the Roman losses.

After the battle of Aix against the Teutons, Marius surprised the Teutons from behind. There was frightful carnage; one hundred thousand Teutons and three hundred Romans killed. [18]

In Sulla's battle of Chaeronea against Archelaus, a general of Mithridates, Sulla had about thirty thousand men, Archelaus, one hundred and ten thousand. Archelaus was beaten by being surprised from the rear. The Romans lost fourteen men, and killed their enemies until worn out in pursuit.

The battle of Orchomenus, against Archelaus, was a repetition of Chaeronea.

Caesar states that his cavalry could not fight the Britons without greatly exposing itself, because they pretended flight in order to get the cavalry away from the infantry and then, dashing from their chariots, they fought on foot with advantage.

A little less than two hundred veterans embarked on a boat which they ran aground at night so as not to be taken by superior naval forces. They reached an advantageous position and passed the night. At the break of day, Otacilius dispatched some four hundred horsemen and some infantry from the Alesio garrison against them. They defended themselves bravely; and having killed some, they rejoined Caesar's troops without having lost a single man.

In Macedonia Caesar's rear-guard was caught by Pompey's cavalry at the passage of the Genusus River, the banks of which were quite steep. Caesar opposed Pompey's cavalry five to seven thousand strong, with his cavalry of six hundred to one thousand men, among which he had taken care to intermingle four hundred picked infantrymen. They did their duty so well that, in the combat that followed, they repulsed the enemy, killed many, and fell back upon their own army without the loss of a single man.

In the battle of Thapsus in Africa, against Scipio, Caesar killed ten thousand, lost fifty, and had some wounded.

In the battle under the walls of Munda in Spain, against one of Pompey's sons, Caesar had eighty cohorts and eight thousand horsemen, about forty-eight thousand men. Pompey with thirteen legions had sixty thousand troops of the line, six thousand cavalry, six thousand light infantry, six thousand auxiliaries; in all, about eighty thousand men. The struggle, says the narrator, was valiantly kept up, step by step, sword to sword. [19]

In that battle of exceptional fury, which hung for a long time in the balance, Caesar had one thousand dead, five hundred wounded; Pompey thirty-three thousand dead, and if Munda had not been so near, scarcely two miles away, his losses would have been doubled. The defensive works of Munda were constructed from dead bodies and abandoned arms.

In studying ancient combats, it can be seen that it was almost always an attack from the flank or rear, a surprise action, that won battles, especially against the Romans. It was in this way that their excellent tactics might be confused. Roman tactics were so excellent that a Roman general who was only half as good as his adversary was sure to be victorious. By surprise alone they could be conquered. Note Xanthippe,—Hannibal—the unexpected fighting methods of the Gauls, etc.

Indeed Xenophon says somewhere, "Be it agreeable or terrible, the less anything is foreseen, the more does it cause pleasure or dismay. This is nowhere better illustrated than in war where every surprise strikes terror even to those who are much the stronger."

But very few fighters armed with cuirass and shield were killed in the front lines.

Hannibal in his victories lost almost nobody but Gauls, his cannon-fodder, who fought with poor shields and without armor.

Nearly always driven in, they fought, nevertheless, with a tenacity that they never showed under any other command.

Thucydides characterizes the combat of the lightly armed, by saying: "As a rule, the lightly armed of both sides took to flight." [20]

In combat with closed ranks there was mutual pressure but little loss, the men not being at liberty to strike in their own way and with all their force.

Caesar against the Nervii, saw his men, who in the midst of the action had instinctively closed in mass in order to resist the mass of barbarians, giving way under pressure. He therefore ordered his ranks and files to open, so that his legionaries, closed in mass, paralyzed and forced to give way to a very strong pressure, might be able to kill and consequently demoralize the enemy. And indeed, as soon as a man in the front rank of the Nervii fell under the blows of the legionaries, there was a halt, a falling back. Following an attack from the rear, and a mêlée, the defeat of the Nervii ensued. [21]

CHAPTER V

MORALE IN ANCIENT BATTLE

We now know the morale and mechanism of ancient fighting; the word mêlée employed by the ancients was many times stronger than the idea to be expressed; it meant a crossing of arms, not a confusion of men.

The results of battles, such as losses, suffice to demonstrate this, and an instant of reflection makes us see the error of the word mêlée. In pursuit it was possible to plunge into the midst of the fugitives, but in combat every one had too much need for the next man, for his neighbor, who was guarding his flanks and his back, to let himself be killed out of sheer wantonness by a sure blow from within the ranks of the enemy. [22]

In the confusion of a real mêlée, Caesar at Pharsalus, and Hannibal at Cannae, would have been conquered. Their shallow ranks, penetrated by the enemy, would have had to fight two against one, they would even have been taken in rear in consequence of the breaking of their ranks.

Also has there not been seen, in troops equally reliable and desperate, that mutual weariness which brings about, with tacit accord, falling back for a breathing spell on both sides in order again to take up the battle?

How can this be possible with a mêlée?

With the confusion and medley of combatants, there might be a mutual extermination, but there would not be any victors. How would they recognize each other? Can you conceive two mixed masses of men or groups, where every one occupied in front can be struck with impunity from the side or from behind? That is mutual extermination, where victory belongs only to survivors; for in the mix-up and confusion, no one can flee, no one knows where to flee.

After all, are not the losses we have seen on both sides demonstration that there was no real mêlée?

The word is, therefore, too strong; the imagination of painters' and poets' has created the mêlée.

This is what happened:

At a charging distance troops marched towards the enemy with all the speed compatible with the necessity for fencing and mutual aid. Quite often, the moral impulse, that resolution to go to the end, manifested itself at once in the order and freedom of gait. That impulse alone put to flight a less resolute adversary.

It was customary among good troops to have a clash, but not the blind and headlong onset of the mass; the preoccupation [23] of the rank was very great, as the behavior of Caesar's troops at Pharsalus shows in their slow march, timed by the flutes of Lacedaemonian battalions. At the moment of getting close to the enemy, the dash slackened of its own accord, because the men of the first rank, of necessity and instinctively, assured themselves of the position of their supports, their neighbors in the same line, their comrades in the second, and collected themselves together in order to be more the masters of their movements to strike and parry. There was a contact of man with man; each took the adversary in front of him and attacked him, because by penetrating into the ranks before having struck him down, he risked being wounded in the side by losing his flank supports. Each one then hit his man with his shield, expecting to make him lose his equilibrium, and at the instant he tried to recover himself landed the blow. The men in the second line, back of the intervals necessary for fencing in the first, were ready to protect their sides against any one that advanced between them and were prepared to relieve tired warriors. It was the same in the third line, and so on.

Every one being supported on either side, the first encounter was rarely decisive, and the fencing, the real combat at close quarters, began.

If men of the first line were wounded quickly, if the other ranks were not in a hurry to relieve or replace them, or if there was hesitation, defeat followed. This happened to the Romans in their first encounters with the Gauls. The Gaul, with his shield, parried the first thrust, brought his big iron sword swooping down with fury upon the top of the Roman shield, split it and went after the man. The Romans, already hesitating before the moral impulse of the Gauls, their ferocious yells, their nudeness, an indication of a contempt for wounds, fell then in a greater number than their adversaries and demoralization followed. Soon they accustomed themselves to this valorous but not tenacious spirit of their enemies, and when they had protected the top of their shields with an iron band, they no longer fell, and the rôles were changed.

The Gauls, in fact, were unable either to hold their ground against the better arms and the thrusts of the Romans, or against their individual superior tenacity, increased nearly tenfold by the possible relay of eight ranks of the maniple. The maniples were self-renewing. Whereas with the Gauls the duration of the combat was limited to the strength of a single man, on account of the difficulties of close or tumultuous ranks, and the impossibility of replacing losses when they were fighting at close quarters.

If the weapons were nearly alike, preserving ranks and thereby breaking down, driving back and confusing the ranks of the enemy, was to conquer. The man in disordered, broken lines, no longer felt himself supported, but vulnerable everywhere, and he fled. It is true that it is hardly possible to break hostile lines without doing the same with one's own. But the one who breaks through first, has been able to do so only by making the foe fall back before his blows, by killing or wounding. He has thereby raised his courage and that of his neighbor. He knows, he sees where he is marching; whilst the adversary overtaken as a consequence of the retreat or the fall of the troops that were flanking him, is surprised. He sees himself exposed on the flank. He falls back on a line with the rank in rear in order to regain support. But the lines in the rear give way to the retreat of the first. If the withdrawal has a certain duration, terror comes as a result of the blows which drive back and mow down the first line. If, to make room for those pushed back, the last lines turn their backs, there is small chance that they will face the front again. Space has tempted them. They will not return to the fight.

Then by that natural instinct of the soldier to worry, to assure himself of his supports, the contagion of flight spreads from the last ranks to the first. The first, closely engaged, has been held to the fight in the meantime, under pain of immediate death. There is no need to explain what follows; it is butchery. (Caedes).

But to return to combat.

It is evident that the formation of troops in a straight line, drawn close together, existed scarcely an instant. Moreover each group of files formed in action was connected with the next group; the groups, like the individuals, were always concerned about their support. The fight took place along the line of contact of the first ranks of the army, a straight line, broken, curved, and bent in different directions according to the various chances of the action at such or such a point, but always restricting and separating the combatants of the two sides. Once engaged on that line, it was necessary to face the front under pain of immediate death. Naturally and necessarily every one in these first ranks exerted all his energy to defend his life.

At no point did the line become entangled as long as there was fighting, for, general or soldier, the effort of each one was to keep up the continuity of support all along the line, and to break or cut that of the enemy, because victory then followed.

We see then that between men armed with swords, it was possible to have, and there was, if the combat was serious, penetration of one mass into the other, but never confusion, or a jumble of ranks, by the men forming these masses. [24]

Sword to sword combat was the most deadly. It presented the most sudden changes, because it was the one in which the individual valor and dexterity of the combatant had the greatest and most immediate influence. Other methods of combat were simpler.

Let us compare pikes and broadswords.

The close formation of men armed with pikes was irresistible so long as it was maintained. A forest of pikes fifteen to eighteen feet long kept you at a distance. [25] On the other hand it was easy to kill off the cavalry and light infantry about the phalanx, which was an unwieldy mass marching with a measured step, and which a mobile body of troops could always avoid. Openings in the phalanx might be occasioned by marching, by the terrain, by the thousand accidents of struggle, by the individual assault of brave men, by the wounded on the ground creeping under the high held pikes and cutting at the legs of the front rank. Men in the phalanx could scarcely see and even the first two lines hardly had a free position for striking. The men were armed with long lances, useless at close quarters, good only for combat at shaft's length (Polybius). They were struck with impunity by the groups [26] which threw themselves into the intervals. And then, once the enemy was in the body of the phalanx, morale disappeared and it became a mass without order, a flock of panic-stricken sheep falling over each other.

In a mob hard-pressed men prick with their knives those who press them. The contagion of fear changes the direction of the human wave; it bends back upon itself and breaks to escape danger. If, then, the enemy fled before the phalanx there was no mêlée. If he gave way tactically before it and availing himself of gaps penetrated it by groups, still there was no mêlée or mixture of ranks. The wedge entering into a mass does not become intermingled with it.

With a phalanx armed with long pikes against a similar phalanx there was still less confusion. They were able to stand for a long time, if the one did not take the other in flank or in rear by a detached body of troops. In all ancient combat, even in victory achieved by methods which affected the morale, such methods are always effective, for man does not change.

It is unnecessary to repeat that in ancient conflicts, demoralization and flight began in the rear ranks.

We have tried to analyze the fight of infantry of the line because its action alone was decisive in ancient combat. The light infantry of both sides took to flight, as Thucydides states. They returned later to pursue and massacre the vanquished. [27]

In cavalry against cavalry, the moral effect of a mass charging in good order was of the greatest influence. We rarely see two cavalry organizations, neither of which breaks before such reciprocal action. Such action was seen on the Tecinus and at Cannae, engagements cited merely because they are very rare exceptions. And even in these cases there was no shock at full speed, but a halt face to face and then an engagement.

The hurricanes of cavalry of those days were poetic figures. They had no reality. In an encounter at full speed, men and horses would be crushed, and neither men nor horses wished such an encounter. The hands of the cavalrymen reined back, the instinct of men and horses was to slacken, to stop, if the enemy himself did not stop, and to make an about if he continued to advance. And if ever they met, the encounter was so weakened by the hands of the men, the rearing of the horses, the swinging of heads, that it was a face to face stop. Some blows were exchanged with the sword or the lance, but the equilibrium was too unstable, mutual support too uncertain for real sword play. Man felt himself too isolated. The moral pressure was too strong. Although not deadly, the combat lasted but a second, precisely because man felt himself, saw himself, alone and surrounded. The first men, who believed themselves no longer supported, could no longer endure uneasiness: they wheeled about and the rest followed. Unless the enemy had also turned, he then pursued at his pleasure until checked by other cavalry, which pursued him in turn.

There never was an encounter between cavalry and infantry. The cavalry harassed with its arrows, with the lance perhaps, while passing rapidly, but it never attacked.

Close conflict on horseback did not exist. And to be sure, if the horse by adding so much to the mobility of man gave him the means of menacing and charging with swiftness, it permitted him to escape with like rapidity when his menace did not shake the enemy. Man by using the horse, pursuant to his natural inclination and sane reasoning, could do as much damage as possible while risking the least possible. To riders without stirrups or saddle, for whom the throwing of the javelin was a difficult matter (Xenophon), combat was but a succession of reciprocal harassings, demonstrations, menaces, skirmishes with arrows. Each cavalry sought an opportunity to surprise, to intimidate, to avail itself of disorder, and to pursue either the cavalry or the infantry. Then "vae victis;" the sword worked.

Man always has had the greatest fear of being trampled upon by horses. That fear has certainly routed a hundred thousand times more men than the real encounter. This was always more or less avoided by the horse, and no one was knocked down. When two ancient cavalry forces wanted really to fight, were forced to it, they fought on foot (Note the Tecinus, Cannae, examples of Livy). I find but little real fighting on horseback in all antiquity like that of Alexander the Great at the passage of the Granicus. Was even that fighting? His cavalry which traversed a river with steep banks defended by the enemy, lost eighty-five men; the Persian cavalry one thousand; and both were equally well armed!

The fighting of the Middle Ages revived the ancient battles except in science. Cavalrymen attacked each other perhaps more than the ancient cavalry did, for the reason that they were invulnerable: it was not sufficient to throw them down; it was necessary to kill when once they were on the ground. They knew, however, that their fighting on horseback was not important so far as results were concerned, for when they wished really to battle, they fought on foot. (Note the combat of the Thirty, Bayard, etc.)

The victors, arrayed in iron from head to foot, lost no one, the peasants did not count. If the vanquished was taken, he was not massacred, because chivalry had established a fraternity of arms between noblemen, the mounted warriors of different nations, and ransom replaced death.

If we have spoken especially of the infantry fight, it is because it was the most serious. On foot, on horseback, on the bridge of a vessel, at the moment of danger, the same man is always found. Any one who knows him well, deduces from his action in the past what his action will be in the future.

CHAPTER VI

UNDER WHAT CONDITIONS REAL COMBATANTS ARE OBTAINED AND HOW THE FIGHTING OF OUR DAYS, IN ORDER TO BE WELL DONE, REQUIRES THEM TO BE MORE DEPENDABLE THAN IN ANCIENT COMBAT

Let us repeat now, what we said at the beginning of this study. Man does not enter battle to fight, but for victory. He does everything that he can to avoid the first and obtain the second. The continued improvement of all appliances of war has no other goal than the annihilation of the enemy. Absolute bravery, which does not refuse battle even on unequal terms, trusting only to God or to destiny, is not natural in man; it is the result of moral culture. It is infinitely rare, because in the face of danger the animal sense of self-preservation always gains the upper hand. Man calculates his chances, with what errors we are about to see.

Now, man has a horror of death. In the bravest, a great sense of duty, which they alone are capable of understanding and living up to, is paramount. But the mass always cowers at sight of the phantom, death. Discipline is for the purpose of dominating that horror by a still greater horror, that of punishment or disgrace. But there always comes an instant when natural horror gets an upper hand over discipline, and the fighter flees. "Stop, stop, hold out a few minutes, an instant more, and you are victor! You are not even wounded yet,—if you turn your back you are dead!" He does not hear, he cannot hear any more. He is full of fear. How many armies have sworn to conquer or perish? How many have kept their oaths? An oath of sheep to stand up against wolves. History shows, not armies, but firm souls who have fought unto death, and the devotion of Thermopylae is therefore justly immortal.

Here we are again brought to the consideration of essential truths, enunciated by many men, now forgotten or unknown.

To insure success in the rude test of conflict, it is not sufficient to have a mass composed of valiant men like the Gauls or the Germans.

The mass needs, and we give it, leaders who have the firmness and decision of command proceeding from habit and an entire faith in their unquestionable right to command as established by tradition, law and society.

We add good arms. We add methods of fighting suitable to these arms and those of the enemy and which do not overtax the physical and moral forces of man. We add also a rational decentralization that permits the direction and employment of the efforts of all even to the last man.

We animate with passion, a violent desire for independence, a religious fanaticism, national pride, a love of glory, a madness for possession. An iron discipline, which permits no one to escape action, secures the greatest unity from top to bottom, between all the elements, between the commanding officers, between the commanding officers and men, between the soldiers.

Have we then a solid army? Not yet. Unity, that first and supreme force of armies, is sought by enacting severe laws of discipline supported by powerful passions. But to order discipline is not enough. A vigilance from which no one may escape in combat should assure the maintenance of discipline. Discipline itself depends on moral pressure which actuates men to advance from sentiments of fear or pride. But it depends also on surveillance, the mutual supervision of groups of men who know each other well.

A wise organization insures that the personnel of combat groups changes as little as possible, so that comrades in peace time maneuvers shall be comrades in war. From living together, and obeying the same chiefs, from commanding the same men, from sharing fatigue and rest, from coöperation among men who quickly understand each other in the execution of warlike movements, may be bred brotherhood, professional knowledge, sentiment, above all unity. The duty of obedience, the right of imposing discipline and the impossibility of escaping from it, would naturally follow.

And now confidence appears.

It is not that enthusiastic and thoughtless confidence of tumultuous or unprepared armies which goes up to the danger point and vanishes rapidly, giving way to a contrary sentiment, which sees treason everywhere. It is that intimate confidence, firm and conscious, which does not forget itself in the heat of action and which alone makes true combatants.

Then we have an army; and it is no longer difficult to explain how men carried away by passions, even men who know how to die without flinching, without turning pale, really strong in the presence of death, but without discipline, without solid organization, are vanquished by others individually less valiant, but firmly, jointly and severally combined.

One loves to picture an armed mob upsetting all obstacles and carried away by a blast of passion.

There is more imagination than truth in that picture. If the struggle depended on individuals, the courageous, impassioned men, composing the mob would have more chance of victory. But in any body of troops, in front of the enemy, every one understands that the task is not the work of one alone, that to complete it requires team work. With his comrades in danger brought together under unknown leaders, he feels the lack of union, and asks himself if he can count on them. A thought of mistrust leads to hesitation. A moment of it will kill the offensive spirit.

Unity and confidence cannot be improvised. They alone can create that mutual trust, that feeling of force which gives courage and daring. Courage, that is the temporary domination of will over instinct, brings about victory.

Unity alone then produces fighters. But, as in everything, there are degrees of unity. Let us see whether modern is in this respect less exacting than ancient combat.

In ancient combat there was danger only at close quarters. If the troops had enough morale (which Asiatic hordes seldom had) to meet the enemy at broadsword's length, there was an engagement. Whoever was that close knew that he would be killed if he turned his back; because, as we have seen, the victors lost but few and the vanquished were exterminated. This simple reasoning held the men and made them fight, if it was but for an instant.

Neglecting the exceptional and very rare circumstances, which may bring two forces together, action to-day is brought on and fought out from afar. Danger begins at great distances, and it is necessary to advance for a long time under fire which at each step becomes heavier. The vanquished loses prisoners, but often, in dead and in wounded, he does not lose more than the victor.

Ancient combat was fought in groups close together, within a small space, in open ground, in full view of one another, without the deafening noise of present day arms. Men in formation marched into an action that took place on the spot and did not carry them thousands of feet away from the starting point. The surveillance of the leaders was easy, individual weakness was immediately checked. General consternation alone caused flight.

To-day fighting is done over immense spaces, along thinly drawn out lines broken every instant by the accidents and the obstacles of the terrain. From the time the action begins, as soon as there are rifle shots, the men spread out as skirmishers or, lost in the inevitable disorder of a rapid march, [28] escape the supervision of their commanding officers. A considerable number conceal themselves; [29] they get away from the engagement and diminish by just so much the material and moral effect and confidence of the brave ones who remain. This can bring about defeat.

But let us look at man himself in ancient combat and in modern. In ancient combat:—I am strong, apt, vigorous, trained, full of calmness, presence of mind; I have good offensive and defensive weapons and trustworthy companions of long standing. They do not let me be overwhelmed without aiding me. I with them, they with me, we are invincible, even invulnerable. We have fought twenty battles and not one of us remained on the field. It is necessary to support each other in time; we see it clearly; we are quick to replace ourselves, to put a fresh combatant in front of a fatigued adversary. We are the legions of Marius, fifty thousand who have held out against the furious avalanches of the Cimbri. We have killed one hundred and forty thousand, taken prisoner sixty thousand, while losing but two or three hundred of our inexperienced soldiers.

To-day, as strong, firm, trained, and courageous as I am, I can never say; I shall return. I have no longer to do with men, whom I do not fear, I have to do with fate in the form of iron and lead. Death is in the air, invisible and blind, whispering, whistling. As brave, good, trustworthy, and devoted as my companions may be, they do not shield me. Only,—and this is abstract and less immediately intelligible to all than the material support of ancient combat,—only I imagine that the more numerous we are who run a dangerous risk, the greater is the chance for each to escape therefrom. I also know that, if we have that confidence which none of us should lack in action, we feel, and we are, stronger. We begin more resolutely, are ready to keep up the struggle longer, and therefore finish it more quickly.

We finish it! But in order to finish it, it is necessary to advance, to attack the enemy, [30] and infantryman or troopers, we are naked against iron, naked against lead, which cannot miss at close range. Let us advance in any case, resolutely. Our adversary will not stand at the point-blank range of our rifle, for the attack is never mutual, we are sure of that. We have been told so a thousand times. We have seen it. But what if matters should change now! Suppose the enemy stands at point-blank range! What of that?

How far this is from Roman confidence!

In another place we have shown that in ancient times to retire from action was both a difficult and perilous matter for the soldier. To-day the temptation is much stronger, the facility greater and the peril less.

Now, therefore, combat exacts more moral cohesion, greater unity than previously. A last remark on the difficulty of obtaining it will complete the demonstration.

Since the invention of fire arms, the musket, the rifle, the cannon, the distances of mutual aid and support have increased among the different arms. [31]

Besides, the facility of communications of all kinds permits the assembling on a given territory of enormous forces. For these reasons, as we have stated, battle fields have become immense.

Supervision becomes more and more difficult. Direction being more distant tends more often to escape from the supreme commanders and the subordinate leaders. The certain and inevitable disorder, which a body of troops always presents in action, is with the moral effect of modern appliances, becoming greater every day. In the midst of the confusion and the vacillation of firing lines, men and commanding officers often lose each other.

Troops immediately and hotly engaged, such as companies and squads, can maintain themselves only if they are well-organized and serve as supports or rallying points to those out of place. Battles tend to become now, more than they have ever been, the battles of men.

This ought not to be true! Perhaps. But the fact is that it is true.

Not all troops are immediately or hotly engaged in battle. Commanding officers always try to keep in hand, as long as possible, some troops capable of marching, acting at any moment, in any direction. To-day, like yesterday, like to-morrow, the decisive action is that of formed troops. Victory belongs to the commander who has known how to keep them in good order, to hold them, and to direct them.

That is incontrovertible.

But commanders can hold out decisive reserves only if the enemy has been forced to commit his.

In troops which do the fighting, the men and the officers closest to them, from corporal to battalion commander, have a more independent action than ever. As it is alone the vigor of that action, more independent than ever of the direction of higher commanders, which leaves in the hands of higher commanders available forces which can be directed at a decisive moment, that action becomes more preponderant than ever. Battles, now more than ever, are battles of men, of captains. They always have been in fact, since in the last analysis the execution belongs to the man in ranks. But the influence of the latter on the final result is greater than formerly. From that comes the maxim of to-day: The battles of men.

Outside of the regulations on tactics and discipline, there is an evident necessity for combating the hazardous predominance of the action of the soldier over that of the commander. It is necessary to delay as long as possible, that instant which modern conditions tend to hasten—the instant when the soldier gets from under the control of the commander.

This completes the demonstration of the truth stated before: Combat requires to-day, in order to give the best results, a moral cohesion, a unity more binding than at any other time. [32] It is as true as it is clear, that, if one does not wish bonds to break, one must make them elastic in order to strengthen them.

CHAPTER VII

PURPOSE OF THIS STUDY

WHAT WOULD BE NECESSARY TO COMPLETE IT

Any other deductions on this subject must come from the meditations of the reader. To be of value in actual application such deductions should be based upon study of modern combat, and that study cannot be made from the accounts of historians alone.

The latter show the action of troop units only in a general way. Action in detail and the individual action of the soldier remain enveloped in a cloud of dust, in narratives as in reality. Yet these questions must be studied, for the conditions they reveal should be the basis of all fighting methods, past, present and future.

Where can data on these questions be found?

We have very few records portraying action as clearly as the report on the engagement at the Pont de l'Hôpital by Colonel Bugeaud. Such stories in even greater detail, for the smallest detail has its importance, secured from participants and witnesses who knew how to see and knew how to remember, are what is necessary in a study of the battle of to-day.

The number of killed, the kind and the character of wounds, often tell more than the longest accounts. Sometimes they contradict them. We want to know how man in general and the Frenchman in particular fought yesterday. Under the pressure of danger, impelled by the instinct for self-preservation, did he follow, make light of, or forget the methods prescribed or recommended? Did he fight in the manner imposed upon him, or in that indicated to him by his instinct or by his knowledge of warfare?

When we have the answers to these questions we shall be very near to knowing how he will conduct himself to-morrow, with and against appliances far more destructive to-day than those of yesterday. Even now, knowing that man is capable only of a given quantity of terror, knowing that the moral effect of destruction is in proportion to the force applied, we are able to predict that, to-morrow less than ever will studied methods be practicable. Such methods are born of the illusions of the field of fire and are opposed to the teachings of our own experience. To-morrow, more than ever, will the individual valor of the soldier and of small groups, be predominant. This valor is secured by discipline.

The study of the past alone can give us a true perception of practical methods, and enable us to see how the soldier will inevitably fight to-morrow.

So instructed, so informed, we shall not be confused; because we shall be able to prescribe beforehand such methods of fighting, such organization, such dispositions as are seen to be inevitable. Such prescriptions may even serve to regulate the inevitable. At any rate they will serve to reduce the element of chance by enabling the commanding officer to retain control as long as possible, and by releasing the individual only at the moment when instinct dominates him.

This is the only way to preserve discipline, which has a tendency to go to pieces by tactical disobedience at the moment of greatest necessity.

It should be understood that the prescriptions in question have to do with dispositions before action; with methods of fighting, and not with maneuvers.

Maneuvers are the movements of troops in the theater of action, and they are the swift and ordered movement on the scene of action of tactical units of all sizes. They do not constitute action. Action follows them.

Confusion in many minds between maneuvers and action brings about doubt and mistrust of our regulation drills. These are good, very good as far as they go, inasmuch as they give methods of executing all movements, of taking all possible formations with rapidity and good order.

To change them, to discuss them, does not advance the question one bit. They do not affect the problem of positive action. Its solution lies in the study of what took place yesterday, from which, alone, it is possible to deduce what will happen to-morrow.

This study must be made, and its result set forth. Each leader, whose worth and authority has been tested in war and recognized by armies, has done something of the sort. Of each of these even might be said, "He knew the soldier; he knew how to make use of him."

The Romans, too, had this knowledge. They obtained it from continuous experience and profound reflexion thereon.

Experience is not continuous to-day. It must be carefully gathered. Study of it should be careful and the results should stimulate reflexion, especially in men of experience. Extremes meet in many things. In ancient times at the point of the pike and sword, armies have conquered similar armies twice their size. Who knows if, in these days of perfected long-range arms of destruction, a small force might not secure, by a happy combination of good sense or genius with morale and appliances, these same heroic victories over a greater force similarly armed?[33]

In spite of the statements of Napoleon I, his assumption that victory is always on the side of the strongest battalions was costly.

PART II

MODERN BATTLE

CHAPTER I

GENERAL DISCUSSION

1. Ancient and Modern Battle

I have heard philosophers reproached for studying too exclusively man in general and neglecting the race, the country, the era, so that their studies of him offer little of real social or political value. The opposite criticism can be made of military men of all countries. They are always eager to expound traditional tactics and organization suitable to the particular character of their race, always the bravest of all races. They fail to consider as a factor in the problem, man confronted by danger. Facts are incredibly different from all theories. Perhaps in this time of military reorganization it would not be out of place to make a study of man in battle and of battle itself.

The art of war is subjected to many modifications by industrial and scientific progress. But one thing does not change, the heart of man. In the last analysis, success in battle is a matter of morale. In all matters which pertain to an army, organization, discipline and tactics, the human heart in the supreme moment of battle is the basic factor. It is rarely taken into account; and often strange errors are the result. Witness the carbine, an accurate and long range weapon, which has never given the service expected of it, because it was used mechanically without considering the human heart. We must consider it!

With improvement in weapons, the power of destruction increases, the moral effect of such weapons increases, and courage to face them becomes rarer. Man does not, cannot change. What should increase with the power of material is the strength of organization, the unity of the fighting machine. Yet these are most neglected. A million men at maneuvers are useless, if a sane and reasoned organization does not assure their discipline, and thereby their reliability, that is, their courage in action.

Four brave men who do not know each other will not dare to attack a lion. Four less brave, but knowing each other well, sure of their reliability and consequently of mutual aid, will attack resolutely. There is the science of the organization of armies in a nutshell.

At any time a new invention may assure victory. Granted. But practicable weapons are not invented every day, and nations quickly put themselves on the same footing as regards armament. The determining factor, leaving aside generals of genius, and luck, is the quality of troops, that is, the organization that best assures their esprit, their reliability, their confidence, their unity. Troops, in this sense, means soldiers. Soldiers, no matter how well drilled, who are assembled haphazard into companies and battalions will never have, have never had, that entire unity which is born of mutual acquaintanceship.

In studying ancient battle, we have seen what a terrible thing battle is. We have seen that man will not really fight except under disciplinary pressure. Even before having studied modern battle, we know that the only real armies are those to which a well thought out and rational organization gives unity throughout battle. The destructive power of improved firearms becomes greater. Battle becomes more open, hindering supervision, passing beyond the vision of the commander and even of subordinate officers. In the same degree, unity should be strengthened. The organization which assures unity of the combatants should be better thought out and more rational. The power of arms increases, man and his weaknesses remain the same. What good is an army of two hundred thousand men of whom only one-half really fight, while the other one hundred thousand disappear in a hundred ways? Better to have one hundred thousand who can be counted upon.

The purpose of discipline is to make men fight in spite of themselves. No army is worthy of the name without discipline. There is no army at all without organization, and all organization is defective which neglects any means to strengthen the unity of combatants. Methods cannot be identical. Draconian discipline does not fit our customs. Discipline must be a state of mind, a social institution based on the salient virtues and defects of the nation.

Discipline cannot be secured or created in a day. It is an institution, a tradition. The commander must have confidence in his right to command. He must be accustomed to command and proud to command. This is what strengthens discipline in armies commanded by an aristocracy in certain countries.

The Prussians do not neglect the homogeneity and consequent unity of organization. They recognize its value. Hessian regiments are composed, the first year, of one-third Hessians, two-thirds Prussians, to control the racial tendencies of troops of a recently annexed country; the second year, of two-thirds Hessians, one-third Prussians; the third year, all Hessians with their own officers.

The Americans have shown us what happens in modern battle to large armies without cohesion. With them the lack of discipline and organization has had the inevitable result. Battle has been between hidden skirmishers, at long distance, and has lasted for days, until some faulty movement, perhaps a moral exhaustion, has caused one or the other of the opposing forces to give way.

In this American War, the mêlées of Agincourt are said to have reappeared, which merely means a mêlée of fugitives. But less than ever has there been close combat.

To fight from a distance is instinctive in man. From the first day he has worked to this end, and he continues to do so. It was thought that with long range weapons close combat might return. On the contrary troops keep further off before its effects.

The primitive man, the Arab, is instability incarnate. A breath, a nothing, governs him at each instant in war. The civilized man, in war, which is opposed to civilization, returns naturally to his first instincts.

With the Arab war remains a matter of agility and cunning. Hunting is his principal pastime and the pursuit of wild beasts teaches the pursuit of man. General Daumas depicts Arabs as cavaliers. What more chivalrous warfare than the night surprise and sack of a camp! Empty words!!

It is commonly said that modern war is the most recondite of things, requiring experts. War, so long as man risks his skin in it, will always be a matter of instinct.

Ancient battle resembled drill. There is no such resemblance in modern battle. This greatly disconcerts both officers and soldiers.

Ancient battles were picnics, for the victors, who lost nobody. Not so today.

Artillery played no part in ancient battle.

The invention of firearms has diminished losses in battle. The improvement of firearms continues to diminish losses. This looks like a paradox. But statistics prove it. Nor is it unreasonable.

Does war become deadlier with the improvement of weapons? Not at all. Man is capable of standing before a certain amount of terror; beyond that he flees from battle. The battle of Pharsalus lasted some four hours. Caesar broke his camp, which is done in the morning; then the formation for battle; then the battle, etc. And he says that his troops were tired, the battle having lasted up to noon. This indicates that he considered it long.

For the middle ages, consult Froissart. The knights in the Battle of the Thirty were armed for battle on foot which they preferred in a serious affair, that is to say in a restricted space. There was a halt, a rest in the combat, when the two parties became exhausted. The Bretons, at this rest, were twenty-five against thirty. The battle had lasted up to exhaustion without loss by the English! Without Montauban the battle would have been terminated by complete and mutual exhaustion and without further losses. For the greater the fatigue, the less strength remained for piercing the armor. Montauban was at the same time felon and hero; felon because he did a thing not permitted by the code of combat; hero, because, if the Bretons had not ably profited by the disorder, he would have been killed when he entered the English formation alone. At the end of the contest the Bretons had four killed, the English eight. Four of the killed were overcome by their armor.

Explain how, under Turenne, men held much longer under fire than to-day. It is perfectly simple. Man is capable of standing before only a certain amount of terror. To-day there must be swallowed in five minutes what took an hour under Turenne. An example will be given.

With the present arms, whose usage is generally known, the instruction of the soldier is of little importance. It does not make the soldier. Take as an example the case of the peasants of the Vendée. Their unity and not individual instruction made them soldiers, whose value could not be denied. Such unity was natural in people of the same village of the same commune, led in battle by their own lords, their own priests, etc.

The greater the perfection of weapons, the more dreadful becomes modern battle, and discipline becomes more difficult to maintain.

The less mobile the troops, the deadlier are battles. Bayonet attacks are not so easily made to-day, and morale consequently is less affected, man fearing man more than death. Astonishing losses seem to have been suffered without breaking by Turenne's armies. Were the casualty reports submitted by the captains of those days correct?

Frederick liked to say that three men behind the enemy were worth more than fifty in front of him, for moral effect. The field of action to-day is more extensive than in Frederick's time. Battle is delivered on more accidented terrain, as armies with great mobility do not need any particular terrain to fight on.

The nature of ancient arms required close order. Modern arms require open order, and they are at the same time of such terrible power that against them too often discipline is broken. What is the solution? Have your combatants opened out? Have them well acquainted with each other so as to have unity. Have reserves to threaten with, held with an iron hand.

Modern weapons have a terrible effect and are almost unbearable by the nervous system. Who can say that he has not been frightened in battle? Discipline in battle becomes the more necessary as the ranks become more open, and the material cohesion of the ranks not giving confidence, it must spring from a knowledge of comrades, and a trust in officers, who must always be present and seen. What man to-day advances with the confidence that rigid discipline and pride in himself gave the Roman soldier, even though the contest is no longer with man but with fate?

To-day the artillery is effective at great distances. There is much liberty of movement for the different arms. The apparent liaison between arms is lessened. This has its influence on morale. There is another advantage in reliable troops, in that they can be extended more widely, and will consequently suffer smaller losses and be in better morale for close conflict.

The further off one is, the more difficult it is to judge of the terrain. Consequently the greater is the necessity for scouting, for reconnoitering the terrain by skirmishers. This is something that the Duke of Gramont forgot at Nordlingen, and which is often forgotten; but it constitutes another important reason for the use of skirmishers.

The formation in rank is a disciplinary measure against the weakness of man in the face of danger. This weakness is greater to-day in that the moral action of weapons is more powerful, and that the material rank has the inherent lack of cohesion of open order. However, open order is necessary to economize losses and permit the use of weapons. Thus to-day there is greater necessity than ever for the rank, that is for discipline, not for the geometrical rank. It is at the same time more necessary and doubly difficult to attain.

In ancient battle unity existed, at least with the Greeks and the Romans. The soldier was known to his officer and comrades; they saw that he fought.

In modern armies where losses are as great for the victor as for the vanquished, the soldier must more often be replaced. In ancient battle the victor had no losses. To-day the soldier is often unknown to his comrades. He is lost in the smoke, the dispersion, the confusion of battle. He seems to fight alone. Unity is no longer insured by mutual surveillance. A man falls, and disappears. Who knows whether it was a bullet or the fear of advancing further that struck him! The ancient combatant was never struck by an invisible weapon and could not fall in this way. The more difficult surveillance, the more necessary becomes the individuality of companies, sections, squads. Not the least of their boasts should be their ability to stand a roll call at all times.

The ancients often avoided hand to hand conflict, so terrible were its consequences. In modern combat, there never is hand to hand conflict if one stands fast.

From day to day close combat tends to disappear. It is replaced by fire action; above all by the moral action of maneuvers. Dispersion brings us back to the necessity for the unity which was an absolute necessity in ancient battle.

Strategy is a game. The first strategist, long before Napoleon, was Horace with his three enemies.

The size of the battle field permits, less than ever, holding units together; the rôle of the general is much more difficult: many more chances are left to fate. Thus the greater the necessity for the best troops who know best their trade, who are most dependable and of greatest fortitude. To diminish the effect of luck, it is necessary to hold longer, to wait for help from a distance. Battles resolve themselves into battles of soldiers. The final decision is more difficult to obtain. There is a strange similarity in battle at one league to battle at two paces. The value of the soldier is the essential element of success. Let us strengthen the soldier by unity.

Battle has more importance than ever. Communication facilities such as the telegraph, concentration facilities such as the railroad, render more difficult such strategic surprises as Ulm and Jena. The whole forces of a country can thus be united. So united, defeat becomes irreparable, disorganization greater and more rapid.

In modern combat the mêlée really exists more than in ancient battle. This appears paradoxical. It is true nevertheless of the mêlée taken in the sense of a mixed up affair where it is infinitely difficult to see clearly.

Man, in the combat of our days, is a man who, hardly knowing how to swim, is suddenly thrown into the sea.

The good quality of troops will more than ever secure victory.

As to the comparative value of troops with cohesion and of new troops, look at the Zouaves of the Guard or the Grenadiers at Magenta, and the 55th at Solferino. [34]

Nothing should be neglected to make the battle order stronger, man stronger.

2. Moral Elements in Battle

When, in complete security, after dinner, in full physical and moral contentment, men consider war and battle they are animated by a noble ardor that has nothing in common with reality. How many of them, however, even at that moment, would be ready to risk their lives? But oblige them to march for days and weeks to arrive at the battle ground, and on the day of battle oblige them to wait minutes, hours, to deliver it. If they were honest they would testify how much the physical fatigue and the mental anguish that precede action have lowered their morale, how much less eager to fight they are than a month before, when they arose from the table in a generous mood.

Man's heart is as changeable as fortune. Man shrinks back, apprehends danger in any effort in which he does not foresee success. There are some isolated characters of an iron temper, who resist the tendency; but they are carried away by the great majority (Bismarck).

Examples show that if a withdrawal is forced, the army is discouraged and takes flight (Frederick). The brave heart does not change.

Real bravery, inspired by devotion to duty, does not know panic and is always the same. The bravery sprung from hot blood pleases the Frenchman more. He understands it, it appeals to his vanity; it is a characteristic of his nature. But it is passing; it fails him at times, especially when there is nothing for him to gain in doing his duty.

The Turks are full of ardor in the advance. They carry their officers with them. But they retreat with the same facility, abandoning their officers.

Mediocre troops like to be led by their shepherds. Reliable troops like to be directed, with their directors alongside of them or behind. With the former the general must be the leader on horseback; with the latter, the manager.

Warnery did not like officers to head a charge. He thought it useless to have them killed before the others. He did not place them in front and his cavalry was good.

General Leboeuf did not favor the proposed advance into battle with platoon leaders in front of the center of their platoons. The fear exists that the fall of the captain will demoralize the rest. What is the solution? Leboeuf must have known that if the officer is not in front of his command, it will advance less confidently, that, with us, all officers are almost always in advance. Practice is stronger than any theory. Therefore fit theories to it. In column, put the chiefs of platoon on the flank where they can see clearly.

Frightfulness! Witness the Turks in the Polish wars. What gave power to the Turks in their wars with Poland was not so much their real strength as their ferocity. They massacred all who resisted; they massacred without the excuse of resistance. Terror preceded them, breaking down the courage of their enemies. The necessity to win or to submit to extreme peril brought about cowardice and submission, for fear of being conquered.

Turenne said, "You tremble, body...." The instinct of self-preservation can then make the strongest tremble. But they are strong enough to overcome their emotion, the fear of advancing, without even losing their heads or their coolness. Fear with them never becomes terror; it is forgotten in the activities of command. He who does not feel strong enough to keep his heart from ever being gripped by terror, should never think of becoming an officer.

The soldiers themselves have emotion. The sense of duty, discipline, pride, the example of their officers and above all their coolness, sustain them and prevent their fear from becoming terror. Their emotion never allows them to sight, or to more than approximately adjust their fire. Often they fire into the air. Cromwell knew this very well, dependable as his troops were, when he said, "Put your trust in God and aim at their shoe laces."

What is too true is that bravery often does not at all exclude cowardice, horrible devices to secure personal safety, infamous conduct.

The Romans were not mighty men, but men of discipline and obstinacy. We have no idea of the Roman military mind, so entirely different from ours. A Roman general who had as little coolness as we have would have been lost. We have incentives in decorations and medals that would have made a Roman soldier run the gauntlet.

How many men before a lion, have the courage to look him in the face, to think of and put into practice measures of self-defense? In war when terror has seized you, as experience has shown it often does, you are as before a lion. You fly trembling and let yourself be eaten up. Are there so few really brave men among so many soldiers? Alas, yes! Gideon was lucky to find three hundred in thirty thousand.

Napoleon said, "Two Mamelukes held three Frenchmen; but one hundred French cavalry did not fear the same number of Mamelukes; three hundred vanquished the same number; one thousand French beat fifteen hundred Mamelukes. Such was the influence of tactics, order and maneuver." In ordinary language, such was the great moral influence of unity, established by discipline and made possible and effective in battle by organization and mutual support. With unity and sensible formation men of an individual value one-third less beat those who were individually their betters. That is the essential, must be the essential, point in the organization of an army. On reflection, this simple statement of Napoleon's seems to contain the whole of battle morale. Make the enemy believe that support is lacking; isolate; cut off, flank, turn, in a thousand ways make his men believe themselves isolated. Isolate in like manner his squadrons, battalions, brigades and divisions; and victory is yours. If, on account of bad organization, he does not anticipate mutual support, there is no need of such maneuver; the attack is enough.

Some men, such as Orientals, Chinese, Tartars, Mongols do not fear death. They are resigned to it at all times. Why is it that they can not stand before the armies of the western people? It is lack of organization. The instinct of self-preservation which at the last moment dominates them utterly, is not opposed by discipline. We have often seen fanatic eastern peoples, implicitly believing that death in battle means a happy and glorious resurrection, superior in numbers, give way before discipline. If attacked confidently, they are crushed by their own weight. In close combat the dagger is better than the bayonet, but instinct is too strong for such people.

What makes the soldier capable of obedience and direction in action, is the sense of discipline. This includes: respect for and confidence in his chiefs; confidence in his comrades and fear of their reproaches and retaliation if he abandons them in danger; his desire to go where others do without trembling more than they; in a word, the whole of esprit de corps. Organization only can produce these characteristics. Four men equal a lion.

Note the army organizations and tactical formations on paper are always determined from the mechanical point of view, neglecting the essential coefficient, that of morale. They are almost always wrong.

Esprit de corps is secured in war. But war becomes shorter and shorter and more and more violent. Consequently, secure esprit de corps in advance.

Mental acquaintanceship is not enough to make a good organization. A good general esprit is needed. All must work for battle and not merely live, quietly going through with drills without understanding their application. Once a man knows how to use his weapon and obey all commands there is needed only occasional drill to brush up those who have forgotten. Marches and battle maneuvers are what is needed.

The technical training of the soldier is not the most difficult. It is necessary for him to know how to use and take care of his weapon; to know how to move to the right and to the left, forward, to the rear, at command, to charge and to march with full pack. But this does not make the soldier. The Vendeans, who knew little of this, were tough soldiers.

It is absolutely necessary to change the instruction, to reduce it to the necessary minimum and to cut out all the superfluities with which peacetime laborers overload it each year. To know the essential well is better than having some knowledge of a lot of things, many of them useless. Teach this the first year, that the second, but the essential from the beginning! Also instruction should be simple to avoid the mental fatigue of long drills that disgust everybody.

Here is a significant sentence in Colonel Borbstaed's enumeration of the reasons for Prussian victory over the Austrians in 1866, "It was ... because each man, being trained, knew how to act promptly and confidently in all phases of battle." This is a fact.

To be held in a building, at every minute of the day to have every movement, every attitude under a not too intelligent surveillance is indeed to be harried. This incessant surveillance weakens the morale of both the watched and the watcher. What is the reason for this incessant surveillance which has long since exceeded shipboard surveillance? Was not that strict enough?

3. Material and Moral Effect

The effect of an army, of one organization on another, is at the same time material and moral. The material effect of an organization is in its power to destroy, the moral effect in the fear that it inspires.

In battle, two moral forces, even more than two material forces, are in conflict. The stronger conquers. The victor has often lost by fire more than the vanquished. Moral effect does not come entirely from destructive power, real and effective as it may be. It comes, above all, from its presumed, threatening power, present in the form of reserves threatening to renew the battle, of troops that appear on the flank, even of a determined frontal attack.

Material effect is greater as instruments are better (weapons, mounts, etc.), as the men know better how to use them, and as the men are more numerous and stronger, so that in case of success they can carry on longer.

With equal or even inferior power of destruction he will win who has the resolution to advance, who by his formations and maneuvers can continually threaten his adversary with a new phase of material action, who, in a word has the moral ascendancy. Moral effect inspires fear. Fear must be changed to terror in order to vanquish.

When confidence is placed in superiority of material means, valuable as they are against an enemy at a distance, it may be betrayed by the actions of the enemy. If he closes with you in spite of your superiority in means of destruction, the morale of the enemy mounts with the loss of your confidence. His morale dominates yours. You flee. Entrenched troops give way in this manner.

At Pharsalus, Pompey and his army counted on a cavalry corps turning and taking Caesar in the rear. In addition Pompey's army was twice as numerous. Caesar parried the blow, and his enemy, who saw the failure of the means of action he counted on, was demoralized, beaten, lost fifteen thousand men put to the sword (while Caesar lost only two hundred) and as many prisoners.

Even by advancing you affect the morale of the enemy. But your object is to dominate him and make him retreat before your ascendancy, and it is certain that everything that diminishes the enemy's morale adds to your resolution in advancing. Adopt then a formation which permits your destructive agency, your skirmishers, to help you throughout by their material action and to this degree diminish that of the enemy.

Armor, in diminishing the material effect that can be suffered, diminishes the dominating moral effect of fear. It is easy to understand how much armor adds to the moral effect of cavalry action, at the critical moment. You feel that thanks to his armor the enemy will succeed in getting to you.

It is to be noted that when a body actually awaits the attack of another up to bayonet distance (something extraordinarily rare), and the attacking troop does not falter, the first does not defend itself. This is the massacre of ancient battle.

Against unimaginative men, who retain some coolness and consequently the faculty of reasoning in danger, moral effect will be as material effect. The mere act of attack does not completely succeed against such troops. (Witness battles in Spain and Waterloo). It is necessary to destroy them, and we are better at this than they by our aptitude in the use of skirmishers and above all in the mad dash of our cavalry. But the cavalry must not be treated, until it comes to so consider itself, as a precious jewel which must be guarded against injury. There should be little of it, but it must be good.

"Seek and ye shall find" not the ideal but the best method that exists. In maneuvers skirmishers, who have some effect, are returned to ranks to execute fire in two ranks which never killed anybody. Why not put your skirmishers in advance? Why sound trumpet calls which they neither hear nor understand? That they do not is fortunate, for each captain has a different call sounded. Example: at Alma, the retreat, etc. [35]

The great superiority of Roman tactics lay in their constant endeavor to coördinate physical and moral effect. Moral effect passes; finally one sees that the enemy is not so terrible as he appeared to be. Physical effect does not. The Greeks tried to dominate. The Romans preferred to kill, and kill they did. They followed thereby the better method. Their moral effect was aided by their reliable and deadly swords.

What moral force is worth to a nation at war is shown by examples. Pichegru played the traitor; this had great influence at home and we were beaten. Napoleon came back; victory returned with him.

But at that we can do nothing without good troops, not even with a Napoleon. Witness Turenne's army after his death. It remained excellent in spite of conflict between and the inefficiency of its two leaders. Note the defensive retreat across the Rhine; the regiment in Champagne attacked in front by infantry and taken in the rear by cavalry. One of the prettiest feats of the art of war.

In modern battle, which is delivered with combatants so far apart, man has come to have a horror of man. He comes to hand to hand fighting only to defend his body or if forced to it by some fortuitous encounter. More than that! It may be said that he seeks to catch the fugitive only for fear that he will turn and fight.

Guilbert says that shock actions are infinitely rare. Here, infinity is taken in its exact mathematical sense. Guilbert reduces to nothing, by deductions from practical examples, the mathematical theory of the shock of one massed body on another. Indeed the physical impulse is nothing. The moral impulse which estimates the attacker is everything. The moral impulse lies in the perception by the enemy of the resolution that animates you. They say that the battle of Amstetten was the only one in which a line actually waited for the shock of another line charging with the bayonets. Even then the Russians gave way before the moral and not before the physical impulse. They were already disconcerted, wavering, worried, hesitant, vacillating, when the blow fell. They waited long enough to receive bayonet thrusts, even blows with the rifle (in the back, as at Inkermann). [36]

This done, they fled. He who calm and strong of heart awaits his enemy, has all the advantage of fire. But the moral impulse of the assailant demoralizes the assailed. He is frightened; he sets his sight no longer; he does not even aim his piece. His lines are broken without defense, unless indeed his cavalry, waiting halted, horsemen a meter apart and in two ranks, does not break first and destroy all formation.

With good troops on both sides, if an attack is not prepared, there is every reason to believe that it will fail. The attacking troops suffer more, materially, than the defenders. The latter are in better order, fresh, while the assailants are in disorder and already have suffered a loss of morale under a certain amount of punishment. The moral superiority given by the offensive movement may be more than compensated by the good order and integrity of the defenders, when the assailants have suffered losses. The slightest reaction by the defense may demoralize the attack. This is the secret of the success of the British infantry in Spain, and not their fire by rank, which was as ineffective with them as with us.

The more confidence one has in his methods of attack or defense, the more disconcerted he is to see them at some time incapable of stopping the enemy. The effect of the present improved fire arm is still limited, with the present organization and use of riflemen, to point blank ranges. It follows that bayonet charges (where bayonet thrusts never occur), otherwise attacks under fire, will have an increasing value, and that victory will be his who secures most order and determined dash. With these two qualities, too much neglected with us, with willingness, with intelligence enough to keep a firm hold on troops in immediate support, we may hope to take and to hold what we take. Do not then neglect destructive effort before using moral effect. Use skirmishers up to the last moment. Otherwise no attack can succeed. It is true it is haphazard fire, nevertheless it is effective because of its volume.

This moral effect must be a terrible thing. A body advances to meet another. The defender has only to remain calm, ready to aim, each man pitted against a man before him. The attacking body comes within deadly range. Whether or not it halts to fire, it will be a target for the other body which awaits it, calm, ready, sure of its effect. The whole first rank of the assailant falls, smashed. The remainder, little encouraged by their reception, disperse automatically or before the least indication of an advance on them. Is this what happens? Not at all! The moral effect of the assault worries the defenders. They fire in the air if at all. They disperse immediately before the assailants who are even encouraged by this fire now that it is over. It quickens them in order to avoid a second salvo.

It is said by those who fought them in Spain and at Waterloo that the British are capable of the necessary coolness. I doubt it nevertheless. After firing, they made swift attacks. If they had not, they might have fled. Anyhow the English are stolid folks, with little imagination, who try to be logical in all things. The French with their nervous irritability, their lively imagination, are incapable of such a defense.

Anybody who thinks that he could stand under a second fire is a man without any idea of battle. (Prince de Ligne).

Modern history furnishes us with no examples of stonewall troops who can neither be shaken nor driven back, who stand patiently the heaviest fire, yet who retire precipitately when the general orders the retreat. (Bismarck).

Cavalry maneuvers, like those of infantry, are threats. The most threatening win. The formation in ranks is a threat, and more than a threat. A force engaged is out of the hand of its commander. I know, I see what it does, what it is capable of. It acts; I can estimate the effect of its action. But a force in formation is in hand; I know it is there, I see it, feel it. It may be used in any direction. I feel instinctively that it alone can surely reach me, take me on the right, on the left, throw itself into a gap, turn me. It troubles me, threatens me. Where is the threatened blow going to fall?

The formation in ranks is a serious threat, which may at any moment be put into effect. It awes one in a terrible fashion. In the heat of battle, formed troops do more to secure victory than do those actively engaged. This is true, whether such a body actually exists or whether it exists only in the imagination of the enemy. In an indecisive battle, he wins who can show, and merely show, battalions and squadrons in hand. They inspire the fear of the unknown.

From the taking of the entrenchments at Fribourg up to the engagement at the bridge of Arcola, up to Solferino, there occur a multitude of deeds of valor, of positions taken by frontal attack, which deceive every one, generals as well as civilians, and which always cause the same mistakes to be made. It is time to teach these folks that the entrenchments at Fribourg were not won by frontal attack, nor was the bridge of Arcola (see the correspondence of Napoleon I), nor was Solferino.

Lieutenant Hercule took fifty cavalry through Alpon, ten kilometers on the flank of the Austrians at Arcola, and the position that held us up for three days, was evacuated. The evacuation was the result of strategic, if not of tactical, moral effect. General or soldier, man is the same.

Demonstrations should be made at greater or less distance, according to the morale of the enemy. That is to say, battle methods vary with the enemy, and an appropriate method should be employed in each individual case.

We have treated and shall treat only of the infantryman. In ancient as in modern battle, he is the one who suffers most. In ancient battle, if he is defeated, he remains because of his slowness at the mercy of the victor. In modern battle the mounted man moves swiftly through danger, the infantryman has to walk. He even has to halt in danger, often and for long periods of time. He who knows the morale of the infantryman, which is put to the hardest proof, knows the morale of all the combatants.

4. The Theory of Strong Battalions

To-day, numbers are considered the essential. Napoleon had this tendency (note his strength reports). The Romans did not pay so much attention to it. What they paid most attention to was to seeing that everybody fought. We assume that all the personnel present with an army, with a division, with a regiment on the day of battle, fights. Right there is the error.

The theory of strong battalions is a shameful theory. It does not reckon on courage but on the amount of human flesh. It is a reflection on the soul. Great and small orators, all who speak of military matters to-day, talk only of masses. War is waged by enormous masses, etc. In the masses, man as an individual disappears, the number only is seen. Quality is forgotten, and yet to-day as always, quality alone produces real effect. The Prussians conquered at Sadowa with made soldiers, united, accustomed to discipline. Such soldiers can be made in three or four years now, for the material training of the soldier is not indeed so difficult.

Caesar had legions that he found unseasoned, not yet dependable, which had been formed for nine years.

Austria was beaten because her troops were of poor quality, because they were conscripts.

Our projected organization will give us four hundred thousand good soldiers. But all our reserves will be without cohesion, if they are thrown into this or that organization on the eve of battle. At a distance, numbers of troops without cohesion may be impressive, but close up they are reduced to fifty or twenty-five per cent. who really fight. Wagram was not too well executed. It illustrated desperate efforts that had for once a moral effect on an impressionable enemy. But for once only. Would they succeed again?

The Cimbrians gave an example [37] and man has not changed. Who to-day is braver than they were? And they did not have to face artillery, nor rifles.

Originally Napoleon found as an instrument, an army with good battle methods, and in his best battles, combat followed these methods. He himself prescribed, at least so they say, for he misrepresented at Saint Helena, the methods used at Wagram, at Eylau, at Waterloo, and engaged enormous masses of infantry which did not give material effect. But it involved a frightful loss of men and a disorder that, after they had once been unleashed, did not permit of the rallying and reemployment that day of the troops engaged. This was a barbaric method, according to the Romans, amateurish, if we may say such a thing of such a man; a method which could not be used against experienced and well trained troops such as d'Erlon's corps at Waterloo. It proved disastrous.

Napoleon looked only at the result to be attained. When his impatience, or perhaps the lack of experience and knowledge in his officers and soldiers, forbade his continued use of real attack tactics, he completely sacrificed the material effect of infantry and even that of cavalry to the moral effect of masses. The personnel of his armies was too changing. In ancient battle victory cost much less than with modern armies, and the same soldiers remained longer in ranks. At the end of his campaigns, when he had soldiers sixty years old, Alexander had lost only seven hundred men by the sword. Napoleon's system is more practicable with the Russians, who naturally group together, mass up, but it is not the most effective. Note the mass formation at Inkermann. [38]

What did Napoleon I do? He reduced the rôle of man in battle, and depended instead on formed masses. We have not such magnificent material.

Infantry and cavalry masses showed, toward the end of the Empire, a tactical degeneracy resulting from the wearing down of their elements and the consequent lowering of standards of morale and training. But since the allies had recognized and adopted our methods, Napoleon really had a reason for trying something so old that it was new to secure that surprise which will give victory once. It can give victory only once however, tried again surprise will be lacking. This was sort of a desperate method which Napoleon's supremacy allowed him to adopt when he saw his prestige waning.

When misfortune and lack of cannon fodder oppressed him, Napoleon became again the practical man not blinded by his supremacy. His entire good sense, his genius, overcame the madness to conquer at all price, and we have his campaign of 1814.

General Ambert says: "Without military traditions, almost without a command, these confused masses (the American armies of the Civil War) struck as men struck at Agincourt and Crecy." At Agincourt and Crecy, we struck very little, but were struck a lot. These battles were great slaughters of Frenchmen, by English and other Frenchmen, who did not greatly suffer themselves. In what, except in disorder, did the American battles resemble these butcheries with the knife? The Americans were engaged as skirmishers at a distance of leagues. In seeking a resemblance the general has been carried away by the mania for phrase-making.

Victory is always for the strong battalions. This is true. If sixty determined men can rout a battalion, these sixty must be found. Perhaps only as many will be found as the enemy has battalions (Note Gideon's proportion of three hundred to thirty thousand of one to one hundred.) Perhaps it would be far and away better, under these circumstances, to fight at night.

5. Combat Methods

Ancient battle was fought in a confined space. The commander could see his whole force. Seeing clearly, his account should have been clear, although we note that many of these ancient accounts are obscure and incomplete, and that we have to supplement them. In modern battle nobody knows what goes on or what has gone on, except from results. Narrations cannot enter into details of execution.

It is interesting to compare tales of feats of arms, narrated by the victor (so-called) or the vanquished. It is hard to tell which account is truthful, if either. Mere assurance may carry weight. Military politics may dictate a perversion of the facts for disciplinary, moral or political reasons. (Note Sommo-Sierra.)

It is difficult even to determine losses, the leaders are such consummate liars. Why is this?

It is bewildering to read a French account and then a foreign account of the same event, the facts stated are so entirely different. What is the truth? Only results can reveal it, such results as the losses on both sides. They are really instructive if they can be gotten at.

I believe that under Turenne there was not existent to the same degree a national pride which tended to hide unpleasant truths. The troops in contending armies were often of the same nation.

If national vanity and pride were not so touchy about recent occurrences, still passionately debated, numerous lessons might be drawn from our last wars. Who can speak impartially of Waterloo, or Waterloo so much discussed and with such heat, without being ashamed? Had Waterloo been won, it would not have profited us. Napoleon attempted the impossible, which is beyond even genius. After a terrible fight against English firmness and tenacity, a fight in which we were not able to subdue them, the Prussians appear. We would have done no better had they not appeared, but they did, very conveniently to sustain our pride. They were confronted. Then the rout began. It did not begin in the troops facing the Prussians but in those facing the English, who were exhausted perhaps, but not more so than their enemies. This was the moral effect of an attack on their right, when they had rather expected reinforcements to appear. The right conformed to the retrograde movement. And what a movement it was!

Why do not authorities acknowledge facts and try to formulate combat methods that conform to reality? It would reduce a little the disorder that bothers men not warned of it. They jump perhaps from the frying pan into the fire. I have known two colonels, one of them a very brave man, who said, "Let soldiers alone before the enemy. They know what to do better than you do." This is a fine statement of French confidence! That they know better than you what should be done. Especially in a panic, I suppose!

A long time ago the Prince de Ligne justified battle formations, above all the famous oblique formation. Napoleon decided the question. All discussion of formations is pedantry. But there are moral reasons for the power of the depth formation.

The difference between practice and theory is incredible. A general, who has given directions a thousand times on the battle field, when asked for directions, gives this order, "Go there, Colonel." The colonel, a man of good sense, says, "Will you explain, sir? What point do you want me to guide on? How far should I extend? Is there anybody on my right? On my left?" The general says, "Advance on the enemy, sir. It seems to me that that ought to be enough. What does this hesitation mean?" But my dear general, what are your orders? An officer should know where his command is, and the command itself should know. Space is large. If you do not know where to send your troops, and how to direct them, to make them understand where they are to go, to give them guides if necessary, what sort of general are you?

What is our method for occupying a fortified work, or a line? We have none! Why not adopt that of Marshal Saxe? Ask several generals how they would do it. They will not know.

There is always mad impatience for results, without considering the means. A general's ability lies in judging the best moment for attack and in knowing how to prepare for it. We took Melegnano without artillery, without maneuver, but at what a price! At Waterloo the Hougoumont farm held us up all day, cost us dear and disorganized us into a mad mob, until Napoleon finally sent eight mortars to smash and burn the château. This is what should have been done at the commencement of the general attack.

A rational and ordered method of combat, or if not ordered, known to all, is enough to make good troops, if there is discipline be it understood. The Portuguese infantry in the Spanish War, to whom the English had taught their method of combat, almost rivalled the English infantry. To-day who has formulated method? Who has a traditional method? Ask the generals. No two will agree.

We have a method, a manner rather, that accords with the national tendency, that of skirmishers in large numbers. But this formation is nowhere formulated. Before a campaign it is decried. Properly so, for it degenerates rapidly into a flock of lost sheep. Consequently troops come to the battle field entirely unused to reality. All the leaders, all the officers, are confused and unoriented. This goes so far that often generals are found who have lost their divisions or brigades; staff officers who have lost their generals and their divisions both; and, although this is more easily understood, many company officers who have lost their commands. This is a serious matter, which might cost us dear in a prolonged war in which the enemy gains experience. Let us hope that experience will lead us, not to change the principle, but to modify and form in a practical way our characteristic battle method of escaping by advancing. The brochure of the Prince of Prussia shows that, without having fought us, the Prussians understand our methods.

There are men such as Marshal Bugeaud who are born warriors in character, mental attitude, intelligence and temperament. They recommend and show by example, such as Colonel Bugeaud's battles in 1815 at the Hospital bridge, tactics entirely appropriate to their national and personal characters. Note Wellington and the Duke of York among the English. But the execution of tactics such as Bugeaud's requires officers who resemble their commanders, at least in courage and decisions. All officers are not of such temper. There is need then of prescribed tactics conforming to the national character, which may serve to guide an ordinary officer without requiring him to have the exceptional ability of a Bugeaud. Such prescribed tactics would serve an officer as the perfectly clear and well defined tactics of the Roman legion served the legion commander. The officer could not neglect them without failing in his duty. Of course they will not make him an exceptional leader. But, except in case of utter incapacity they will keep him from entirely failing in his task, from making absurd mistakes. Nor will they prevent officers of Bugeaud's temper from using their ability. They will on the contrary help them by putting under their command men prepared for the details of battle, which will not then come to them as a surprise.

This method need not be as completely dogmatic as the Roman. Our battle is too varying an affair. But some clearly defined rules, established by experience, would prevent the gross errors of inefficients. (Such as causing skirmishers to fall back when the formed rank fires, and consequently allowing them to carry with them in their retreat, the rank itself.) They would be useful aids to men of coolness and decision.

The laying down of such tactics would answer the many who hold that everything is improvised on the battle field and who find no better improvisation than to leave the soldier to himself. (See above.)

We should try to exercise some control over our soldiers, who advance by flight (note the Vendeans) or escape by advancing, as you like. But if something unexpected surprises them, they flee as precipitately.

Invention is less needed than verification, demonstration and organization of proper methods. To verify; observe better. To demonstrate; try out and describe better. To organize, distribute better, bearing in mind that cohesion means discipline. I do not know who put things that way; but it is truer than ever in this day of invention.

With us very few reason or understand reason, very few are cool. Their effect is negligible in the disorder of the mass; it is lost in numbers. It follows that we above all need a method of combat, sanely thought out in advance. It must be based on the fact that we are not passively obedient instruments, but very nervous and restless people, who wish to finish things quickly and to know in advance where we are going. It must be based on the fact that we are very proud people, but people who would all skulk if we were not seen, and who consequently must always be seen, and act in the presence of our comrades and of the officers who supervise us. From this comes the necessity for organizing the infantry company solidly. It is the infantryman on whom the battle has the most violent effect, for he is always most exposed; it is he therefore who must be the most solidly supported. Unity must be secured by a mutual acquaintanceship of long standing between all elements.

If you only use combat methods that require leaders without fear, of high intelligence, full of good sense, of esprit, you will always make mistakes. Bugeaud's method was the best for him. But it is evident, in his fight at the Hospital bridge that his battalion commanders were useless. If he had not been there, all would have been lost. He alone, omnipresent, was capable of resolute blows that the others could not execute. His system can be summed up in two phrases; always attack even when on the defensive; fire and take cover only when not attacked. His method was rational, considering his mentality and the existing conditions, but in carrying it into execution he judged his officers and soldiers by himself and was deceived. No dogmatic principles can be drawn from his method, nor from any other. Man is always man. He does not always possess ability and resolution. The commander must make his choice of methods, depending on his troops and on himself.

The essential of tactics is: the science of making men fight with their maximum energy. This alone can give an organization with which to fight fear. This has always been true.

We must start here and figure mathematically. Mathematics is the dominant science in war, just as battle is its only purpose. Pride generally causes refusal to acknowledge the truth that fear of being vanquished is basic in war. In the mass, pride, vanity, is responsible for this dissimulation. With the tiny number of absolutely fearless men, what is responsible is their ignorance of a thing they do not feel. There is however, no real basis but this, and all real tactics are based on it. Discipline is a part of tactics, is absolutely at the base of tactics, as the Romans showed. They excelled the Gauls in intelligence, but not in bravery.

To start with: take battalions of four companies, four platoons each, in line or in column. The order of battle may be: two platoons deployed as skirmishers, two companies in reserve, under command of the battalion commander. In obtaining a decision destructive action will come from skirmishers. This action should be directed by battalion commanders, but such direction is not customary. No effect will be secured from skirmishers at six hundred paces. They will never, never, never, be nicely aligned in front of their battalions, calm and collected, after an advance. They will not, even at maneuvers. The battalion commander ought to be advanced enough to direct his skirmishers. The whole battalion, one-half engaged, one-half ready for any effort, ought to remain under his command, under his personal direction as far as possible. In the advance the officers, the soldiers, are content if they are merely directed; but, when the battle becomes hot, they must see their commander, know him to be near. It does not matter even if he is without initiative, incapable of giving an order. His presence creates a belief that direction exists, that orders exist, and that is enough.

When the skirmishers meet with resistance, they fall back to the ranks. It is the rôle of reserves to support and reinforce the line, and above all, by a swift charge to cut the enemy's line. This then falls back and the skirmishers go forward again, if the advance is resumed. The second line should be in the formation, battalions in line or in column, that hides it best. Cover the infantry troops before their entry into action; cover them as much as possible and by any means; take advantage of the terrain; make them lie down. This is the English method in defense of heights, instanced in Spain and at Waterloo. Only one bugle to each battalion should sound calls. What else is there to be provided for?

Many haughty generals would scream protests like eagles if it were suggested that they take such precautions for second line battalions or first line troops not committed to action. Yet this is merely a sane measure to insure good order without the slightest implication of cowardice. [39]

With breech-loading weapons, the skirmishers on the defensive fire almost always from a prone position. They are made to rise with difficulty, either for retreat or for advance. This renders the defense more tenacious....

CHAPTER II

INFANTRY

1. Masses—Deep Columns.

Study of the effect of columns brings us to the consideration of mass operations in general. Read this singular argument in favor of attacks by battalions in close columns: "A column cannot stop instantly without a command. Suppose your first rank stops at the instant of shock: the twelve ranks of the battalion, coming up successively, would come in contact with it, pushing it forward.... Experiments made have shown that beyond the sixteenth the impulsion of the ranks in rear has no effect on the front, it is completely taken up by the fifteen ranks already massed behind the first.... To make the experiment, march at charging pace and command halt to the front rank without warning the rest. The ranks will precipitate themselves upon each other unless they be very attentive, or unless, anticipating the command, they check themselves unconsciously while marching."

But in a real charge, all your ranks are attentive, restless, anxious about what is taking place at the front and, if the latter halts, if the first line stops, there will be a movement to the rear and not to the front. Take a good battalion, possessed of extraordinary calmness and coolness, thrown full speed on the enemy, at one hundred and twenty steps to the minute. To-day it would have to advance under a fire of five shots a minute! At this last desperate moment if the front rank stops, it will not be pushed, according to the theory of successive impulses, it will be upset. The second line will arrive only to fall over the first and so on. There should be a drill ground test to see up to what rank this falling of the pasteboard figures would extend.

Physical impulse is merely a word. If the front rank stops it will let itself fall and be trampled under foot rather than cede to the pressure that pushes it forward. Any one experienced in infantry engagements of to-day knows that is just what happens. This shows the error of the theory of physical impulse—a theory that continues to dictate as under the Empire (so strong is routine and prejudice) attacks in close column. Such attacks are marked by absolute disorder and lack of leadership. Take a battalion fresh from barracks, in light marching order; intent only on the maneuver to be executed. It marches in close column in good order; its subdivisions are full four paces apart. The non-commissioned officers control the men. But it is true that if the terrain is slightly accidented, if the guide does not march with mathematical precision, the battalion in close column becomes in the twinkling of an eye a flock of sheep. What would happen to a battalion in such a formation, at one hundred paces from the enemy? Nobody will ever see such an instance in these days of the rifle.

If the battalion has marched resolutely, if it is in good order, it is ten to one that the enemy has already withdrawn without waiting any longer. But suppose the enemy does not flinch? Then the man of our days, naked against iron and lead, no longer controls himself. The instinct of preservation controls him absolutely. There are two ways of avoiding or diminishing the danger; they are to flee or to throw one-self upon it. Let us rush upon it. Now, however small the intervals of space and time that separate us from the enemy, instinct shows itself. We rush forward, but ... generally, we rush with prudence, with a tendency to let the most urgent ones, the most intrepid ones, pass on. It is strange, but true, that the nearer we approach the enemy, the less we are closed up. Adieu to the theory of pressure. If the front rank is stopped, those behind fall down rather than push it. Even if this front rank is pushed, it will itself fall down rather than advance. There is nothing to wonder at, it is sheer fact. Any pushing is to the rear. (Battle of Diernstein.)

To-day more than ever flight begins in the rear, which is affected quite as much as the front.

Mass attacks are incomprehensible. Not one out of ten was ever carried to completion and none of them could be maintained against counter-attacks. They can be explained only by the lack of confidence of the generals in their troops. Napoleon expressly condemns in his memoirs such attacks. He, therefore, never ordered them. But when good troops were used up, and his generals believed they could not obtain from young troops determined attacks in tactical formation, they came back to the mass formation, which belongs to the infancy of the art, as a desperate resort.

If you use this method of pressing, of pushing, your force will disappear as before a magician's wand.

But the enemy does not stand; the moral pressure of danger that precedes you is too strong for him. Otherwise, those who stood and aimed even with empty rifles, would never see a charge come up to them. The first line of the assailant would be sensible of death and no one would wish to be in the first rank. Therefore, the enemy never merely stands; because if he does, it is you that flee. This always does away with the shock. The enemy entertains no smaller anxiety than yours. When he sees you near, for him also the question is whether to flee or to advance. Two moral impulses are in conflict.

This is the instinctive reasoning of the officer and soldier, "If these men wait for me to close with them, it means death. I will kill, but I will undoubtedly be killed. At the muzzle of the gun-barrel the bullet can not fail to find its mark. But if I can frighten them, they will run away. I can shoot them and bayonet in the back. Let us make a try at it." The trial is made, and one of the two forces, at some stage of the advance, perhaps only at two paces, makes an about and gets the bayonet in the back.

Imagination always sees loaded arms and this fancy is catching.

The shock is a mere term. The de Saxe, the Bugeaud theory: "Close with the bayonet and with fire action at close quarters. That is what kills people and the victor is the one who kills most," is not founded on fact. No enemy awaits you if you are determined, and never, never, never, are two equal determinations opposed to each other. It is well known to everybody, to all nations, that the French have never met any one who resisted a bayonet charge.

The English in Spain, marching resolutely in face of the charges of the French in column, have always defeated them.... The English were not dismayed at the mass. If Napoleon had recalled the defeat of the giants of the Armada by the English vessels, he might not have ordered the use of the d'Erlon column.

Blücher in his instructions to his troops, recalled that the French have never held out before the resolute march of the Prussians in attack column....

Suvaroff used no better tactics. Yet his battalions in Italy drove us at the point of their bayonets.

Each nation in Europe says: "No one stands his ground before a bayonet charge made by us." All are right. The French, no more than others, resist a resolute attack. All are persuaded that their attacks are irresistible; that an advance will frighten the enemy into flight. Whether the bayonet be fixed or in the scabbard makes no difference....

There is an old saying that young troops become uneasy if any one comes upon them in a tumult and in disorder; the old troops, on the contrary, see victory therein. At the commencement of a war, all troops are young. Our impetuosity pushes us to the front like fools ... the enemy flees. If the war lasts, everybody becomes inured. The enemy no longer troubles himself when in front of troops charging in a disordered way, because he knows and feels that they are moved as much by fear as by determination. Good order alone impresses the enemy in an attack, for it indicates real determination. That is why it is necessary to secure good order and retain it to the very last. It is unwise to take the running step prematurely, because you become a flock of sheep and leave so many men behind that you will not reach your objective. The close column is absurd; it turns you in advance into a flock of sheep, where officers and men are jumbled together without mutual support. It is then necessary to march as far as possible in such order as best permits the action of the non-commissioned officers, the action of unity, every one marching in front of eye-witnesses, in the open. On the other hand, in closed columns man marches unobserved and on the slightest pretext he lies down or remains behind. Therefore, it is best always to keep the skirmishers in advance or on the flanks, and never to recall them when in proximity to the enemy. To do so establishes a counter current that carries away your men. Let your skirmishers alone. They are your lost children; they will know best how to take care of themselves.

To sum up: there is no shock of infantry on infantry. There is no physical impulse, no force of mass. There is but a moral impulse. No one denies that this moral impulse is stronger as one feels better supported, that it has greater effect on the enemy as it menaces him with more men. From this it follows that the column is more valuable for the attack than the deployed order.

It might be concluded from this long statement that a moral pressure, which always causes flight when a bold attack is made, would not permit any infantry to hold out against a cavalry charge; never, indeed, against a determined charge. But infantry must resist when it is not possible to flee, and until there is complete demoralization, absolute terror, the infantry appreciates this. Every infantryman knows it is folly to flee before cavalry when the rifle is infallible at point-blank, at least from the rider's point of view. It is true that every really bold charge ought to succeed. But whether man is on foot or on horseback, he is always man. While on foot he has but himself to force; on horseback he must force man and beast to march against the enemy. And mounted, to flee is so easy. (Remark by Varney).

We have seen than in an infantry mass those in rear are powerless to push those in front unless the danger is greater in rear. The cavalry has long understood this. It attacks in a column at double distance rather than at half-distance, in order to avoid the frightful confusion of the mass. And yet, the allurement of mathematical reasoning is such that cavalry officers, especially the Germans, have seriously proposed attacking infantry by deep masses, so that the units in rear might give impulse to those in front. They cite the proverb, "One nail drives the other." What can you say to people who talk such nonsense? Nothing, except, "Attack us always in this way."

Real bayonet attacks occurred in the Crimean war. (Inkermann). [40] They were carried out by a small force against a larger one. The power of mass had no influence in such cases. It was the mass which fell back, turned tail even before the shock. The troops who made the bold charge did nothing but strike and fire at backs. These instances show men unexpectedly finding themselves face to face with the enemy, at a distance at which a man can close fearlessly without falling out on the way breathless. They are chance encounters. Man is not yet demoralized by fire; he must strike or fall back.... Combat at close quarters does not exist. At close quarters occurs the ancient carnage when one force strikes the other in the back.

Columns have absolutely but a moral effect. They are threatening dispositions....

The mass impulse of cavalry has long been discredited. You have given up forming it in deep ranks although cavalry possesses a speed that would bring on more of a push upon the front at a halt than the last ranks of the infantry would bring upon the first. Yet you believe in the mass action of infantry!

As long as the ancient masses marched forward, they did not lose a man and no one lay down to avoid the combat. Dash lasted up to the time of stopping; the run was short in every case. In modern masses, in French masses especially, the march can be continued, but the mass loses while marching under fire. Moral pressure, continually exerted during a long advance, stops one-half of the combatants on the way. To-day, above all in France, man protests against such use of his life. The Frenchman wants to fight, to return blow for blow. If he is not allowed to, this is what happens. It happened to Napoleon's masses. Let us take Wagram, where his mass was not repulsed. Out of twenty-two thousand men, three thousand to fifteen hundred reached the position. Certainly the position was not carried by them, but by the material and moral effect of a battery of one hundred pieces, cavalry, etc., etc. Were the nineteen thousand missing men disabled? No. Seven out of twenty-two, a third, an enormous proportion may have been hit. What became of the twelve thousand unaccounted for? They had lain down on the road, had played dummy in order not to go on to the end. In the confused mass of a column of deployed battalions, surveillance, difficult enough in a column at normal distances, is impossible. Nothing is easier than dropping out through inertia; nothing more common.

This thing happens to every body of troops marching forward, under fire, in whatever formation it may be. The number of men falling out in this way, giving up at the least opportunity, is greater as formation is less fixed and the surveillance of officers and comrades more difficult. In a battalion in closed column, this kind of temporary desertion is enormous; one-half of the men drop out on the way. The first platoon is mingled with the fourth. They are really a flock of sheep. No one has control, all being mixed. Even if, in virtue of the first impulse, the position is carried, the disorder is so great that if it is counter-attacked by four men, it is lost.

The condition of morale of such masses is fully described in the battle of Caesar against the Nervii, Marius against the Cimbri. [41]

What better arguments against deep columns could there be than the denials of Napoleon at St. Helena?

2. Skirmishers—Supports—Reserves—Squares

This is singular. The cavalry has definite tactics. Essentially it knows how it fights. The infantry does not.

Our infantry no longer has any battle tactics; the initiative of the soldier rules. The soldiers of the First Empire trusted to the moral and passive action of masses. To-day, the soldiers object to the passive action of masses. They fight as skirmishers, or they march to the front as a flock of sheep of which three-fourths seek cover enroute, if the fire is heavy. The first method, although better than the second, is bad unless iron discipline and studied and practical methods of fighting insure maintaining strong reserves. These should be in the hands of the leaders and officers for support purposes, to guard against panics, and to finish by the moral effect of a march on the enemy, of flank menaces, etc., the destructive action of the skirmishers.

To-day when the ballistic arm is so deadly, so effective, a unit which closes up in order to fight is a unit in which morale is weakened.

Maneuver is possible only with good organization; otherwise it is no more effective than the passive mass or a rabble in an attack.

In ancient combat, the soldier was controlled by the leader in engagements; now that fighting is open, the soldier cannot be controlled. Often he cannot even be directed. Consequently it is necessary to begin an action at the latest possible moment, and to have the immediate commanders understand what is wanted, what their objectives are, etc.

In the modern engagement, the infantryman gets from under our control by scattering, and we say: a soldier's war. Wrong, wrong. To solve this problem, instead of scattering to the winds, let us increase the number of rallying points by solidifying the companies. From them come battalions; from battalions come regiments.

Action in open order was not possible nor evident under Turenne. The majority of the soldiers that composed the army, were not held near at hand, in formation. They fought badly. There was a general seeking for cover. Note the conduct of the Americans in their late war.

The organization of the legion of Marshal Saxe shows the strength of the tendency toward shock action as opposed to fire action.

The drills, parades and firing at Potsdam were not the tactics of Old Fritz. Frederick's secret was promptitude and rapidity of movement. But they were popularly believed to be his means. People were fond of them, and are yet. The Prussians for all their leaning toward parade, mathematics, etc., ended by adopting the best methods. The Prussians of Jena were taken in themselves by Frederick's methods. But since then they have been the first to strike out in a practical way, while we, in France, are still laboring at the Potsdam drills.

The greater number of generals who fought in the last wars, under real battle conditions, ask for skirmishers in large units, well supported. Our men have such a strong tendency to place themselves in such units even against the will of their leaders, that they do not fight otherwise.

A number of respectable authors and military men advocate the use of skirmishers in large bodies, as being dictated by certain necessities of war. Ask them to elucidate this mode of action, and you will see that this talk of skirmishers in large bodies is nothing else but an euphemism for absolute disorder. An attempt has been made to fit the theory to the fact. Yet the use of skirmishers in large bodies is absurd with Frenchmen under fire, when the terrain and the sharpness of the action cause the initiative and direction to escape from the commanders, and leave it to the men, to small groups of soldiers.

Arms are for use. The best disposition for material effect in attack or defense is that which permits the easiest and most deadly use of arms. This disposition is the scattered thin line. The whole of the science of combat lies then in the happy, proper combination, of the open order, scattered to secure destructive effect, and a good disposition of troops in formation as supports and reserves, so as to finish by moral effect the action of the advanced troops. The proper combination varies with the enemy, his morale and the terrain. On the other hand, the thin line can have good order only with a severe discipline, a unity which our men attain from pride. Pride exists only among people who know each other well, who have esprit de corps, and company spirit. There is a necessity for an organization that renders unity possible by creating the real individuality of the company.

Self-esteem is unquestionably one of the most powerful motives which moves our men. They do not wish to pass for cowards in the eyes of their comrades. If they march forward they want to distinguish themselves. After every attack, formation (not the formation of the drill ground but that adopted by those rallying to the chief, those marching with him,) no longer exists. This is because of the inherent disorder of every forward march under fire. The bewildered men, even the officers, have no longer the eyes of their comrades or of their commander upon them, sustaining them. Self-esteem no longer impels them, they do not hold out; the least counter-offensive puts them to rout.

The experience of the evening ought always to serve the day following; but as the next day is never identical with the evening before, the counsel of experience can not be applied to the latter. When confused battalions shot at each other some two hundred paces for some time with arms inferior to those of our days, flight commenced at the wings. Therefore, said experience, let us reënforce the wings, and the battalion was placed between two picked companies. But it was found that the combat methods had been transformed. The elite companies were then reassembled into picked corps and the battalion, weaker than ever, no longer had reënforced wings. Perhaps combat in open order predominates, and the companies of light infantrymen being, above all, skirmishers, the battalion again is no longer supported. In our day the use of deployed battalions as skirmishers is no longer possible; and one of the essential reasons for picked companies is the strengthening of the battalion.

The question has been asked; Who saved the French army on the Beresina and at Hanau? The Guard, it is true. But, outside of the picked corps, what was the French army then? Droves, not troops. Abnormal times, abnormal deeds. The Beresina, Hanau, prove nothing to-day.

With the rapid-firing arms of infantry to-day, the advantage belongs to the defense which is completed by offensive movements carried out at opportune times.

Fire to-day is four or five times more rapid even if quite as haphazard as in the days of muzzle loaders. Everybody says that this renders impossible the charges of cavalry against infantry which has not been completely thrown into disorder, demoralized. What then must happen to charges of infantry, which marches while the cavalry charges?

Attacks in deep masses are no longer seen. They are not wise, and never were wise. To advance to the attack with a line of battalions in column, with large intervals and covered by a thick line of skirmishers, when the artillery has prepared the terrain, is very well. People with common sense have never done otherwise. But the thick line of skirmishers is essential. I believe that is the crux of the matter.

But enough of this. It is simple prudence for the artillery to prepare the infantry action by a moment's conversation with the artillery of the enemy infantry. If that infantry is not commanded by an imbecile, as it sometimes is, it will avoid that particular conversation the arguments of which would break it up, although they may not be directed precisely in its direction. All other things being equal, both infantries suffer the same losses in the artillery duel. The proportion does not vary, however complete the artillery preparation.

One infantry must always close with another under rapid fire from troops in position, and such a fire is, to-day more than ever, to the advantage of the defense. Ten men come towards me; they are at four hundred meters; with the ancient arm, I have time to kill but two before they reach me; with rapid fire, I have time to kill four or five. Morale does not increase with losses. The eight remaining might reach me in the first case; the five or six remaining will certainly not in the second.

If distance be taken, the leader can be seen, the file-closers see, the platoon that follows watches the preceding. Dropping out always exists, but it is less extensive with an open order, the men running more risks of being recognized. Stragglers will be fewer as the companies know each other better, and as the officers and men are more dependable.

It is difficult, if not impossible, to get the French infantry to make use of its fire before charging. If it fires, it will not charge, because it will continue to fire. (Bugeaud's method of firing during the advance is good.) What is needed, then, is skirmishers, who deliver the only effective fire, and troops in formation who push the skirmishers on, in themselves advancing to the attack.

The soldier wants to be occupied, to return shot for shot. Place him in a position to act immediately, individually. Then, whatever he does, you have not wholly lost your authority over him.

Again and again and again, at drill, the officers and non-commissioned officer ought to tell the private: "This is taught you to serve you under such circumstances." Generals, field officers, ought to tell officers the same thing. This alone can make an instructed army like the Roman army. But to-day, who of us can explain page for page, the use of anything ordered by our tactical regulations except the school of the skirmisher? "Forward," "retreat," and "by the flank," are the only practical movements under fire. But the others should be explained. Explain the position of "carry arms" with the left hand. Explain the ordinary step. Explain firing at command in the school of the battalion. It is well enough for the school of the platoon, because a company can make use thereof, but a battalion never can.

Everything leads to the belief that battle with present arms will be, in the same space of time, more deadly than with ancient ones. The trajectory of the projectile reaching further, the rapidity of firing being four times as great, more men will be put out of commission in less time. While the arm becomes more deadly, man does not change, his morale remains capable of certain efforts and the demands upon it become stronger. Morale is overtaxed; it reaches more rapidly the maximum of tension which throws the soldier to the front or rear. The rôle of commanders is to maintain morale, to direct those movements which men instinctively execute when heavily engaged and under the pressure of danger.

Napoleon I said that in battle, the rôle of skirmishers is the most fatiguing and most deadly. This means that under the Empire, as at present, the strongly engaged infantry troops rapidly dissolved into skirmishers. The action was decided by the moral agency of the troops not engaged, held in hand, capable of movement in any direction and acting as a great menace of new danger to the adversary, already shaken by the destructive action of the skirmishers. The same is true to-day. But the greater force of fire arms requires, more than ever, that they be utilized. The rôle of the skirmisher becomes preëminently the destructive role; it is forced on every organization seriously engaged by the greater moral pressure of to-day which causes men to scatter sooner.

Commanders-in-chief imagine formed battalions firing on the enemy and do not include the use of skirmishers in drill. This is an error, for they are necessary in drill and everywhere, etc. The formed rank is more difficult to utilize than ever. General Leboeuf used a very practical movement of going into battle, by platoons, which advance to the battle line in echelon, and can fire, even if they are taken in the very act of the movement. There is always the same dangerous tendency toward mass action even for a battalion in maneuver. This is an error. The principles of maneuver for small units should not be confused with those for great units. Emperor Napoleon did not prescribe skirmishers in flat country. But every officer should be reduced who does not utilize them to some degree.

The rôle of the skirmisher becomes more and more predominant. He should be so much the more watched and directed as he is used against more deadly arms, and, consequently, is more disposed to escape from all control, from all direction. Yet under such battle conditions formations are proposed which send skirmishers six hundred paces in advance of battalions and which give the battalion commander the mission of watching and directing (with six companies of one hundred and twenty men) troops spread over a space of three hundred paces by five hundred, at a minimum. To advance skirmishers six hundred paces from their battalion and to expect they will remain there is the work of people who have never observed.

Inasmuch as combat by skirmishers tends to predominate and since it becomes more difficult with the increase of danger, there has been a constant effort to bring into the firing line the man who must direct it. Leaders have been seen to spread an entire battalion in front of an infantry brigade or division so that the skirmishers, placed under a single command, might obey a general direction better. This method, scarcely practicable on the drill-ground, and indicating an absolute lack of practical sense, marks the tendency. The authors of new drills go too far in the opposite direction. They give the immediate command of the skirmishers in each battalion to the battalion commander who must at the same time lead his skirmishers and his battalion. This expedient is more practical than the other. It abandons all thought of an impossible general control and places the special direction in the right hands. But the leadership is too distant, the battalion commander has to attend to the participation of his battalion in the line, or in the ensemble of other battalions of the brigade or division, and the particular performance of his skirmishers. The more difficult, confused, the engagement becomes, the more simple and clear ought to be the roles of each one. Skirmishers are in need of a firmer hand than ever to direct and maintain them, so that they may do their part. The battalion commander must be entirely occupied with the rôle of skirmishers, or with the rôle of the line. There should be smaller battalions, one-half the number in reserve, one-half as skirmisher battalions. In the latter the men should be employed one-half as skirmishers and one-half held in reserve. The line of skirmishers will then gain steadiness.

Let the battalion commander of the troops of the second line entirely occupy himself with his battalion.

The full battalion of six companies is to-day too unwieldy for one man. Have battalions of four companies of one hundred men each, which is certainly quite sufficient considering the power of destruction which these four companies place in the hands of one man. He will have difficulty in maintaining and directing these four companies under the operation of increasingly powerful modern appliances. He will have difficulty in watching them, in modern combat, with the greater interval between the men in line that the use of the present arms necessitates. With a unified battalion of six hundred men, I would do better against a battalion of one thousand Prussians, than with a battalion of eight hundred men, two hundred of whom are immediately taken out of my control.

Skirmishers have a destructive effect; formed troops a moral effect. Drill ground maneuvers should prepare for actual battle. In such maneuvers, why, at the decisive moment of an attack, should you lighten the moral anxiety of the foe by ceasing his destruction, by calling back your skirmishers? If the enemy keeps his own skirmishers and marches resolutely behind them, you are lost, for his moral action upon you is augmented by his destructive action against which you have kindly disarmed yourself.

Why do you call back your skirmishers? Is it because your skirmishers hinder the operation of your columns, block bayonet charges? One must never have been in action to advance such a reason. At the last moment, at the supreme moment when one or two hundred meters separate you from the adversary, there is no longer a line. There is a fearless advance, and your skirmishers are your forlorn hope. Let them charge on their own account. Let them be passed or pushed forward by the mass. Do not recall them. Do not order them to execute any maneuver for they are not capable of any, except perhaps, that of falling back and establishing a counter-current which might drag you along. In these moments, everything hangs by a thread. Is it because your skirmishers would prevent you from delivering fire? Do you, then, believe in firing, especially in firing under the pressure of approaching danger, before the enemy? If he is wise, certainly he marches preceded by skirmishers, who kill men in your ranks and who have the confidence of a first success, of having seen your skirmishers disappear before them. These skirmishers will certainly lie down before your unmasked front. In that formation they easily cause you losses, and you are subjected to their destructive effect and to the moral effect of the advance of troops in formation against you. Your ranks become confused; you do not hold the position. There is but one way of holding it, that is to advance, and for that, it is necessary at all costs to avoid firing before moving ahead. Fire opened, no one advances further.

Do you believe in opening and ceasing fire at the will of the commander as on the drill ground? The commencement of fire by a battalion, with the present arms especially, is the beginning of disorder, the moment where the battalion begins to escape from its leader. While drilling even, the battalion commanders, after a little lively drill, after a march, can no longer control the fire.

Do you object that no one ever gets within two hundred meters of the enemy? That a unit attacking from the front never succeeds? So be it! Let us attack from the flank. But a flank is always more or less covered. Men are stationed there, ready for the blow. It will be necessary to pick off these men.

To-day, more than ever, no rapid, calm firing is possible except skirmish firing.

The rapidity of firing has reduced six ranks to two ranks. With reliable troops who have no need of the moral support of a second rank behind them, one rank suffices to-day. At any rate, it is possible to await attack in two ranks.

In prescribing fire at command, in seeking to minimize the rôle of skirmishers instead of making it predominate, you take sides with the Germans. We are not fitted for that sort of game. If they adopt fire at command, it is just one more reason for our finding another method. We have invented, discovered the skirmisher; he is forced upon us by our men, our arms, etc. He must be organized.

In fire by rank, in battle, men gather into small groups and become confused. The more space they have, the less will be the disorder.

Formed in two ranks, each rank should be still thinner. All the shots of the second line are lost. The men should not touch; they should be far apart. The second rank in firing from position at a supreme moment, ought not to be directly behind the first. The men ought to be echeloned behind the first. There will always be firing from position on any front. It is necessary to make this firing as effective and as easy as possible. I do not wish to challenge the experiences of the target range but I wish to put them to practical use.

It is evident that the present arms are more deadly than the ancient ones; the morale of the troops will therefore be more severely shaken. The influence of the leader should be greater over the combatants, those immediately engaged. If it seems rational, let colonels engage in action, with the battalions of their regiment in two lines. One battalion acts as skirmishers; the other battalion waits, formed ready to aid the first. If you do not wish so to utilize the colonels, put all the battalions of the regiment in the first line, and eventually use them as skirmishers. The thing is inevitable; it will be done in spite of you. Do it yourself at the very first opportunity.

The necessity of replenishing the ammunition supply so quickly used up by the infantry, requires engaging the infantry by units only, which can be relieved by other units after the exhaustion of the ammunition supply. As skirmishers are exhausted quickly, engage entire battalions as skirmishers, assisted by entire battalions as supports or reserves. This is a necessary measure to insure good order. Do not throw into the fight immediately the four companies of the battalion. Up to the crucial moment, the battalion commander ought to guard against throwing every one into the fight.

There is a mania, seen in our maneuver camps, for completely covering a battle front, a defended position, by skirmishers, without the least interval between the skirmishers of different battalions. What will be the result? Initially a waste of men and ammunition. Then, difficulty in replacing them.

Why cover the front everywhere? If you do, then what advantage is there in being able to see from a great distance? Leave large intervals between your deployed companies. We are no longer only one hundred meters from the enemy at the time of firing. Since we are able to see at a great distance we do not risk having the enemy dash into these intervals unexpectedly. Your skirmisher companies at large intervals begin the fight, the killing. While your advance companies move ahead, the battalion commander follows with his formed companies, defilading them as much as possible. He lets them march. If the skirmishers fight at the halt, he supervises them. If the commanding officer wishes to reënforce his line, if he wants to face an enemy who attempts to advance into an interval, if he has any motive for doing it, in a word, he rushes new skirmishers into the interval. Certainly, these companies have more of the forward impulse, more dash, if dash is needed, than the skirmishers already in action. If they pass the first skirmishers, no harm is done. There you have echelons already formed. The skirmishers engaged, seeing aid in front of them, can be launched ahead more easily.

Besides, the companies thrown into this interval are a surprise for the enemy. That is something to be considered, as is the fact that so long as there is fighting at a halt, intervals in the skirmish lines are fit places for enemy bullets. Furthermore, these companies remain in the hands of their leaders. With the present method of reënforcing skirmishers—I am speaking of the practical method of the battlefield, not of theory—a company, starting from behind the skirmishers engaged, without a place in which to deploy, does not find anything better to do than to mingle with the skirmishers. Here it doubles the number of men, but in doing so brings disorder, prevents the control of the commanders and breaks up the regularly constituted groups. While the closing up of intervals to make places for new arrivals is good on the drill ground, or good before or after the combat, it never works during battle.

No prescribed interval will be kept exactly. It will open, it will close, following the fluctuations of the combat. But the onset, during which it can be kept, is not the moment of brisk combat; it is the moment of the engagement, of contact, consequently, of feeling out. It is essential that there remain space in which to advance. Suppose you are on a plain, for in a maneuver one starts from the flat terrain. In extending the new company it will reënforce the wings of the others, the men naturally supporting the flanks of their comrades. The individual intervals will lessen in order to make room for the new company. The company will always have a well determined central group, a rallying point for the others. If the interval has disappeared there is always time to employ the emergency method of doubling the ranks in front; but one must not forget, whatever the course taken, to preserve good order.

We cannot resist closing intervals between battalions; as if we were still in the times of the pikemen when, indeed, it was possible to pass through an interval! To-day, the fighting is done ten times farther away, and the intervals between battalions are not weak joints. They are covered by the fire of the skirmishers, as well covered by fire as the rest of the front, and invisible to the enemy.

Skirmishers and masses are the formations for action of poorly instructed French troops. With instruction and unity there would be skirmishers supported and formation in battalion columns at most.

Troops in close order can have only a moral effect, for the attack, or for a demonstration. If you want to produce a real effect, use musketry. For this it is necessary to form a single line. Formations have purely moral effect. Whoever counts on their material, effective action against reliable, cool troops, is mistaken and is defeated. Skirmishers alone do damage. Picked shots would do more if properly employed.

In attacking a position, start the charge at the latest possible moment, when the leader thinks he can reach the objective not all out of breath. Until then, it has been possible to march in rank, that is under the officers, the rank not being the mathematical line, but the grouping in the hands of the leader, under his eye. With the run comes confusion. Many stop, the fewer as the run is shorter. They lie down on the way and will rejoin only if the attack succeeds, if they join at all. If by running too long the men are obliged to stop in order to breathe and rest, the dash is broken, shattered. At the advance, very few will start. There are ten chances to one of seeing the attack fail, of turning it into a joke, with cries of "Forward with fixed bayonet," but none advancing, except some brave men who will be killed uselessly. The attack vanishes finally before the least demonstration of the foe. An unfortunate shout, a mere nothing, can destroy it.

Absolute rules are foolish, the conduct of every charge being an affair requiring tact. But so regulate by general rules the conduct of an infantry charge that those who commence it too far away can properly be accused of panic. And there is a way. Regulate it as the cavalry charge is regulated, and have a rearguard in each battalion of non-commissioned officers, of most reliable officers, in order to gather together, to follow close upon the charge, at a walk, and to collect all those who have lain down so as not to march or because they were out of breath. This rearguard might consist of a small platoon of picked shots, such as we need in each battalion. The charge ought to be made at a given distance, else it vanishes, evaporates. The leader who commences it too soon either has no head, or does not want to gain his objective.

The infantry of the line, as opposed to elite commands, should not be kept in support. The least firm, the most impressionable, are thus sent into the road stained with the blood of the strongest. We place them, after a moral anxiety of waiting, face to face with the terrible destruction and mutilation of modern weapons. If antiquity had need of solid troops as supports, we have a greater need of them. Death in ancient combat was not as horrible as in the modern battle where the flesh is mangled, slashed by artillery fire. In ancient combat, except in defeat, the wounded were few in number. This is the reply to those who wish to begin an action by chasseurs, zouaves, etc.

He, general or mere captain, who employs every one in the storming of a position can be sure of seeing it retaken by an organized counter-attack of four men and a corporal.

In order that we may have real supervision and responsibility in units from companies to brigades, the supporting troops ought to be of the same company, the same battalion, the same brigade, as the case may be. Each brigade ought to have its two lines, each battalion its skirmishers, etc.

The system of holding out a reserve as long as possible for independent action when the enemy has used his own, ought to be applied downwards. Each battalion should have its own, each regiment its own, firmly maintained.

There is more need than ever to-day, for protecting the supporting forces, the reserves. The power of destruction increases, the morale remains the same. The tests of morale, being more violent than previously, ought to be shorter, because the power of morale has not increased. The masses, reserves, the second, the first lines, should be protected and sheltered even more than the skirmishers.

Squares sometimes are broken by cavalry which pursues the skirmishers into the square. Instead of lying down, they rush blindly to their refuge which they render untenable and destroy. No square can hold out against determined troops.... But!

The infantry square is not a thing of mechanics, of mathematical reasoning; it is a thing of morale. A platoon in four ranks, two facing the front, two the rear, its flanks guarded by the extreme files that face to the flank, and conducted, supported by the non-commissioned officers placed in a fifth rank, in the interior of the rectangle, powerful in its compactness and its fire, cannot be dislodged by cavalry. However, this platoon will prefer to form a part of a large square, it will consider itself stronger, because of numbers, and indeed it will be, since the feeling of force pervades this whole force. This feeling is power in war.

People who calculate only according to the fire delivered, according to the destructive power of infantry, would have it fight deployed against cavalry. They do not consider that although supported and maintained, although such a formation seem to prevent flight, the very impetus of the charge, if led resolutely, will break the deployment before the shock arrives. It is clear that if the charge is badly conducted, whether the infantry be solid or not, it will never reach its objective. Why? Moral reasons and no others make the soldier in a square feel himself stronger than when in line. He feels himself watched from behind and has nowhere to flee.

3. Firing

It is easy to misuse breech-loading weapons, such as the rifle. The fashion to-day is to use small intrenchments, covering battalions. As old as powder. Such shelter is an excellent device on the condition, however, that behind it, a useful fire can be delivered.

Look at these two ranks crouched under the cover of a small trench. Follow the direction of the shots. Even note the trajectory shown by the burst of flame. You will be convinced that, under such conditions, even simple horizontal firing is a fiction. In a second, there will be wild firing on account of the noise, the crowding, the interference of the two ranks. Next everybody tries to get under the best possible cover. Good-by firing.

It is essential to save ammunition, to get all possible efficiency from the arm. Yet the official adoption of fire by rank insures relapsing into useless firing at random. Good shots are wasted, placed where it is impossible for them to fire well.

Since we have a weapon that fires six times more rapidly than the ancient weapon, why not profit by it to cover a given space with six times fewer riflemen than formerly? Riflemen placed at greater intervals, will be less bewildered, will see more clearly, will be better watched (which may seem strange to you), and will consequently deliver a better fire than formerly. Besides, they will expend six times less ammunition. That is the vital point. You must always have ammunition available, that is to say, troops which have not been engaged. Reserves must be held out. This is hard to manage perhaps. It is not so hard to manage, however, as fire by command.

What is the use of fire by rank? By command? It is impracticable against the enemy, except in extraordinary cases. Any attempt at supervision of it is a joke! File firing? The first rank can shoot horizontally, the only thing required; the second rank can fire only into the air. It is useless to fire with our bulky knapsacks interfering so that our men raise the elbow higher than the shoulder. Learn what the field pack can be from the English, Prussians, Austrians, etc.... Could the pack not be thicker and less wide? Have the first rank open; let the second be checkerwise; and let firing against cavalry be the only firing to be executed in line.

One line will be better than two, because it will not be hindered by the one behind it. One kind of fire is practicable and efficient, that of one rank. This is the fire of skirmishers in close formation.

The king's order of June 1st, 1776, reads (p. 28): "Experience in war having proved that three ranks fire standing, and the intention of his majesty being to prescribe only what can be executed in front of the enemy, he orders that in firing, the first man is never to put his knee on the ground, and that the three ranks fire standing at the same time." This same order includes instructions on target practice, etc.

Marshal de Gouvion-Saint Cyr says that conservatively one-fourth of the men who are wounded in an affair are put out of commission by the third rank. This estimate is not high enough if it concerns a unit composed of recruits like those who fought at Lützen and Bautzen. The marshal mentions the astonishment of Napoleon when he saw the great number of men wounded in the hand and forearm. This astonishment of Napoleon's is singular. What ignorance in his marshals not to have explained such wounds! Chief Surgeon Larrey, by observation of the wounds, alone exonerated our soldiers of the accusation of self-inflicted wounds. The observation would have been made sooner, had the wounds heretofore been numerous. That they had not been can be explained only by the fact that while the young soldiers of 1813 kept instinctively close in ranks, up to that time the men must have spaced themselves instinctively, in order to be able to shoot. Or perhaps in 1813, these young men might have been allowed to fire a longer time in order to distract them and keep them in ranks, and not often allowed to act as skirmishers for fear of losing them. Whilst formerly, the fire by rank must have been much rarer and fire action must have given way almost entirely to the use of skirmishers.

Fire by command presupposes an impossible coolness. Had any troops ever possessed it they would have mowed down battalions as one mows down corn stalks. Yet it has been known for a long time, since Frederick, since before Frederick, since the first rifle. Let troops get the range calmly, let them take aim together so that no one disturbs or hinders the other. Have each one see clearly, then, at a signal, let them all fire at once. Who is going to stand against such people? But did they aim in those days? Not so accurately, possibly, but they knew how to shoot waist-high, to shoot at the feet. They knew how to do it. I do not say they did it. If they had done so, there would not have been any need of reminding them of it so often. Note Cromwell's favorite saying, "Aim at their shoe-laces;" that of the officers of the empire, "Aim at the height of the waist." Study of battles, of the expenditure of bullets, show us no such immediate terrible results. If such a means of destruction was so easy to obtain, why did not our illustrious forbears use it and recommend it to us? (Words of de Gouvion-Saint-Cyr.)

Security alone creates calmness under fire.

In minor operations of war, how many captains are capable of tranquilly commanding their fire and maneuvering with calmness?

Here is a singular thing. You hear fire by rank against cavalry seriously recommended in military lectures. Yet not a colonel, not a battalion commander, not a captain, requires this fire to be executed in maneuvers. It is always the soldier who forces the firing. He is ordered to shoot almost before he aims for fear he will shoot without command. Yet he ought to feel that when he is aiming, his finger on the trigger, his shot does not belong to him, but rather to the officer who ought to be able to let him aim for five minutes, if advisable, examining, correcting the positions, etc. He ought, when aiming, always be ready to fire upon the object designated, without ever knowing when it will please his commander to order him to fire.

Fire at command is not practicable in the face of the enemy. If it were, the perfection of its execution would depend on the coolness of the commander and the obedience of the soldier. The soldier is the more easily trained.

The Austrians had fire by command in Italy against cavalry. Did they use it? They fired before the command, an irregular fire, a fire by file, with defective results.

Fire by command is impossible. But why is firing by rank at will impossible, illusory, under the fire of the enemy? Because of the reasons already given and, for this reason: that closed ranks are incompatible with fire-arms, on account of the wounding caused by the latter in ranks. In closed ranks, the two lines touching elbows, a man who falls throws ten men into complete confusion. There is no room for those who drop and, however few fall, the resulting disorder immediately makes of the two ranks a series of small milling groups. If the troops are young, they become a disordered flock before any demonstration. (Caldiero, Duhesme.) If the troops have some steadiness, they of themselves will make space: they will try to make way for the bullets: they will scatter as skirmishers with small intervals. (Note the Grenadier Guards at Magenta.)[42]

With very open ranks, men a pace apart, whoever falls has room, he is noticed by a lesser number, he drags down no one in his fall. The moral impression on his comrades is less. Their courage is less impaired. Besides, with rapid fire everywhere, spaced ranks with no man in front of another, at least permit horizontal fire. Closed ranks permit it hardly in the first rank, whose ears are troubled by the shots from the men behind. When a man has to fire four or five shots a minute, one line is certainly more solid than two, because, while the firing is less by half, it is more than twice as likely to be horizontal fire as in the two-rank formation. Well-sustained fire, even with blank cartridges, would be sufficient to prevent a successful charge. With slow fire, two ranks alone were able to keep up a sufficiently continuous fusillade. With rapid fire, a single line delivers more shots than two with ancient weapons. Such fire, therefore, suffices as a fusillade.

Close ranks, while suitable for marching, do not lend themselves to firing at the halt. Marching, a man likes a comrade at his side. Firing, as if he felt the flesh attracting the lead, he prefers being relatively isolated, with space around him. Breech-loading rifles breed queer ideas. Generals are found who say that rapid firing will bring back fire at command, as if there ever were such a thing. They say it will bring back salvo firing, thus permitting clear vision. As if such a thing were possible! These men have not an atom of common sense.

It is singular to see a man like Guibert, with practical ideas on most things, give a long dissertation to demonstrate that the officers of his time were wrong in aiming at the middle of the body, that is, in firing low. He claims this is ridiculous to one who understands the trajectory of the rifle. These officers were right. They revived the recommendations of Cromwell, because they knew that in combat the soldier naturally fires too high because he does not aim, and because the shape of the rifle, when it is brought to the shoulder, tends to keep the muzzle higher than the breech. Whether that is the reason or something else, the fact is indisputable. It is said that in Prussian drills all the bullets hit the ground at fifty paces. With the arms of that time and the manner of fighting, results would have been magnificent in battle if the bullets had struck fifty paces before the enemy instead of passing over his head.

Yet at Mollwitz, where the Austrians had five thousand men disabled, the Prussians had over four thousand.

Firing with a horizontal sector, if the muzzle be heavy, is more deadly than firing with a vertical sector.

4. Marches. Camps. Night Attacks.

From the fact that infantry ought always to fight in thin formation, scattered, it does not follow that it ought to be kept in that order. Only in column is it possible to maintain the battle order. It is necessary to keep one's men in hand as long as possible, because once engaged, they no longer belong to you.

The disposition in closed mass is not a suitable marching formation, even in a battalion for a short distance. On account of heat, the closed column is intolerable, like an unventilated room. Formation with half-distances is better. (Why? Air, view, etc.)

Such a formation prevents ready entry of the column into battle in case of necessity or surprise. The half-divisions not in the first line are brought up, the arms at the order, and they can furnish either skirmishers or a reserve for the first line which has been deployed as skirmishers.

At Leuctra, Epaminondas diminished, by one-half, the depth of his men; he formed square phalanxes of fifty men to a side. He could have very well dispensed with it, for the Lacedaemonian right was at once thrown into disorder by its own cavalry which was placed in front of that wing. The superior cavalry of Epaminondas overran not only the cavalry but the infantry that was behind it. The infantry of Epaminondas, coming in the wake of his cavalry finished the work. Turning to the right, the left of Epaminondas then took in the flank the Lacedaemonian line. Menaced also in front by the approaching echelons of Epaminondas, this line became demoralized and took to flight. Perhaps this fifty by fifty formation was adopted in order to give, without maneuver, a front of fifty capable of acting in any direction. At Leuctra, it simply acted to the right and took the enemy in the flank and in reverse.

Thick woods are generally passed through in close column. There is never any opening up, with subsequent closing on the far side. The resulting formation is as confused as a flock of sheep.

In a march through mountains, difficult country, a bugler should be on the left, at the orders of an intelligent officer who indicates when the halt seems necessary for discipline in the line. The right responds and if the place has been judged correctly an orderly formation is maintained. Keep in ranks. If one man steps out, others follow. Do not permit men to leave ranks without requiring them to rejoin.

In the rear-guard it is always necessary to have pack mules in an emergency; without this precaution, considerable time may be lost. In certain difficult places time is thus lost every day.

In camp, organize your fatigue parties in advance; send them out in formation and escorted.

Definite and detailed orders ought to be given to the convoy, and the chief baggage-master ought to supervise it, which is rarely the case.

It is a mistake to furnish mules to officers and replace them in case of loss or sickness. The officer overloads the mule and the Government loses more thereby than is generally understood. Convoys are endless owing to overloaded mules and stragglers. If furnished money to buy a mule the officer uses it economically because it is his. If mules are individually furnished to officers instead of money, the officer will care for his beast for the same reason. But it is better to give money only, and the officer, if he is not well cared for on the march has no claim against the Government.

Always, always, take Draconian measures to prevent pillage from commencing. If it begins, it is difficult ever to stop it. A body of infantry is never left alone. There is no reason for calling officers of that arm inapt, when battalions although established in position are not absolutely on the same line, with absolutely equal intervals. Ten moves are made to achieve the exact alignment which the instructions on camp movements prescribe. Yet designating a guiding battalion might answer well enough and still be according to the regulations.

Why are not night attacks more employed to-day, at least on a grand scale? The great front which armies occupy renders their employment more difficult, and exacts of the troops an extreme aptitude in this kind of surprise tactics (found in the Arabs, Turcos, Spahis), or absolute reliability. There are some men whose knowledge of terrain is wonderful, with an unerring eye for distance, who can find their way through places at night which they have visited only in the day time. Utilizing such material for a system of guides it would be possible to move with certainty. These are simple means, rarely employed, for conducting a body of troops into position on the darkest night. There is, even, a means of assuring at night the fire of a gun upon a given point with as much precision as in plain day.

CHAPTER III

CAVALRY

1. Cavalry and Modern Appliances

They say that cavalry is obsolete; that it can be of no use in battles waged with the weapons of today. Is not infantry affected in the same way?

Examples drawn from the last two wars are not conclusive. In a siege, in a country which is cut off, one does not dare to commit the cavalry, and therefore takes from it its boldness, which is almost its only weapon.

The utility of cavalry has always been doubted. That is because its cost is high. It is little used, just because it does cost. The question of economy is vital in peace times. When we set a high value upon certain men, they are not slow to follow suit, and to guard themselves against being broken. Look at staff officers who are almost never broken (reduced), even when their general himself is.

With new weapons the rôle of cavalry has certainly changed less than any other, although it is the one which is most worried about. However, cavalry always has the same doctrine: Charge! To start with, cavalry action against cavalry is always the same. Also against infantry. Cavalry knows well enough today, as it has always known, that it can act only against infantry which has been broken. We must leave aside epic legends that are always false, whether they relate to cavalry or infantry. Infantry cannot say as much of its own action against infantry. In this respect there is a complete anarchy of ideas. There is no infantry doctrine.

With the power of modern weapons, which forces you to slow down if it does not stop you, the advance under fire becomes almost impossible. The advantage is with the defensive. This is so evident that only a madman could dispute it. What then is to be done? Halt, to shoot at random and cannonade at long range until ammunition is exhausted? Perhaps. But what is sure, is that such a state of affairs makes maneuver necessary. There is more need than ever for maneuver at a long distance in an attempt to force the enemy to shift, to quit his position. What maneuver is swifter than that of cavalry? Therein is its role.

The extreme perfection of weapons permits only individual action in combat, that is action by scattered forces. At the same time it permits the effective employment of mass action out of range, of maneuvers on the flank or in the rear of the enemy in force imposing enough to frighten him.

Can the cavalry maneuver on the battle field? Why not? It can maneuver rapidly, and above all beyond the range of infantry fire, if not of artillery fire. Maneuver being a threat, of great moral effect, the cavalry general who knows how to use it, can contribute largely to success. He arrests the enemy in movement, doubtful as to what the cavalry is going to attempt. He makes the enemy take some formation that keeps him under artillery fire for a while, above all that of light artillery if the general knows how to use it. He increases the enemy's demoralization and thus is able to rejoin his command.

Rifled cannon and accurate rifles do not change cavalry tactics at all. These weapons of precision, as the word precision indicates, are effective only when all battle conditions, all conditions of aiming, are ideal. If the necessary condition of suitable range is lacking, effect is lacking. Accuracy of fire at a distance is impossible against a troop in movement, and movement is the essence of cavalry action. Rifled weapons fire on them of course, but they fire on everybody.

In short, cavalry is in the same situation as anybody else.

What response is there to this argument? Since weapons have been improved, does not the infantryman have to march under fire to attack a position? Is the cavalryman not of the same flesh? Has he less heart than the infantryman? If one can march under fire, cannot the other gallop under it?

When the cavalryman cannot gallop under fire, the infantryman cannot march under it. Battles will consist of exchanges of rifle shots by concealed men, at long range. The battle will end only when the ammunition is exhausted.

The cavalryman gallops through danger, the infantryman walks. That is why, if he learns, as it is probable he will, to keep at the proper distance, the cavalryman will never see his battle rôle diminished by the perfection of long range fire. An infantryman will never succeed by himself. The cavalryman will threaten, create diversions, worry, scatter the enemy's fire, often even get to close quarters if he is properly supported. The infantryman will act as usual. But more than ever will he need the aid of cavalry in the attack. He who knows how to use his cavalry with audacity will inevitably be the victor. Even though the cavalryman offers a larger target, long range weapons will paralyze him no more than another.

The most probable effect of artillery of today, will be to increase the scattering in the infantry, and even in the cavalry. The latter can start in skirmisher formation at a distance and close in while advancing, near its objective. It will be more difficult to lead; but this is to the advantage of the Frenchman.

The result of improving the ballistics of the weapon, for the cavalry as for the infantry (there is no reason why it should be otherwise for the cavalry), will be that a man will flee at a greater distance from it, and nothing more.

Since the Empire, the opinion of European armies is that the cavalry has not given the results expected of it.

It has not given great results, for the reason that we and others lacked real cavalry generals. He is, it seems, a phenomenon that is produced only every thousand years, more rarely than a real general of infantry. To be a good general, whether of infantry or cavalry, is an infinitely rare thing, like the good in everything. The profession of a good infantry general is as difficult as, perhaps more difficult than, that of a good cavalry general. Both require calmness. It comes more easily to the cavalryman than to the foot soldier who is much more engaged. Both require a like precision, a judgment of the moral and physical forces of the soldier; and the morale of the infantryman, his constitution, is more tried than is the case with the horseman.

The cavalry general, of necessity, sees less clearly; his vision has its limits. Great cavalry generals are rare. Doubtless Seidlitz could not, in the face of the development of cannon and rifle, repeat his wonders. But there is always room for improvement. I believe there is much room for improvement.

We did not have under the Empire a great cavalry general who knew how to handle masses. The cavalry was used like a blind hammer that strikes heavily and not always accurately. It had immense losses. Like the Gauls, we have a little too much confidence in the "forward, forward, not so many methods." Methods do not hinder the forward movement. They prepare the effect and render it surer and at the same time less costly to the assailant. We have all the Gallic brutality. (Note Marignano, where the force of artillery and the possibility of a turning movement around a village was neglected). What rare things infantry and cavalry generals are!

A leader must combine resolute bravery and impetuosity with prudence and calmness; a difficult matter!

The broken terrain of European fields no longer permits, we are told, the operation of long lines, of great masses of cavalry. I do not regret it. I am struck more with the picturesque effect of these hurricanes of cavalry in the accounts of the Empire than with the results obtained. It does not seem to me that these results were in proportion to the apparent force of the effort and to the real grandeur of the sacrifices. And indeed, these enormous hammers (a usual figure), are hard to handle. They have not the sure direction of a weapon well in hand. If the blow is not true, recovery is impossible, etc. However, the terrain does not to-day permit the assembling of cavalry in great masses. This compelling reason for new methods renders any other reason superfluous.

Nevertheless, the other reasons given in the ministerial observations of 1868, on the cavalry service, seems to me excellent. The improvement of appliances, the extension of battle fields, the confidence to the infantry and the audacity to the artillery that the immediate support of the cavalry gives, demand that this arm be in every division in sufficient force for efficient action.

I, therefore, think it desirable for a cavalry regiment to be at the disposal of a general commanding a division. Whatever the experiences of instruction centers, they can not change in the least my conviction of the merit of this measure in the field.

2. Cavalry Against Cavalry

Cavalry action, more than that of infantry, is an affair of morale.

Let us study first the morale of the cavalry engagement in single combat. Two riders rush at each other. Are they going to direct their horses front against front? Their horses would collide, both would be forced to their feet, while running the chance of being crushed in the clash or in the fall of their mounts. Each one in the combat counts on his strength, on his skill, on the suppleness of his mount, on his personal courage; he does not want a blind encounter, and he is right. They halt face to face, abreast, to fight man to man; or each passes the other, thrusting with the sabre or lance; or each tries to wound the knee of the adversary and dismount him in this way. But as each is trying to strike the other, he thinks of keeping out of the way himself, he does not want a blind encounter that does away with the combat. The ancient battles, the cavalry engagements, the rare cavalry combats of our days, show us nothing else.

Discipline, while keeping the cavalrymen in the ranks, has not been able to change the instinct of the rider. No more than the isolated man is the rider in the line willing to meet the shock of a clash with the enemy. There is a terrible moral effect in a mass moving forward. If there is no way to escape to the right or to the left, men and horses will avoid the clash by stopping face to face. But only preëminently brave troops, equally seasoned in morale, alike well led and swept along, animated alike, will meet face to face. All these conditions are never found united on either side, so the thing is never seen. Forty-nine times out of fifty, one of the cavalry forces will hesitate, bolt, get into disorder, flee before the fixed purpose of the other. Three quarters of the time this will happen at a distance, before they can see each other's eyes. Often they will get closer. But always, always, the stop, the backward movement, the swerving of horses, the confusion, bring about fear or hesitation. They lessen the shock and turn it into instant flight. The resolute assailant does not have to slacken. He has not been able to overcome or turn the obstacles of horses not yet in flight, in this uproar of an impossible about face executed by routed troops, without being in disorder himself. But this disorder is that of victory, of the advance, and a good cavalry does not trouble itself about it. It rallies in advancing, while the vanquished one has fear at its heels.

On the whole, there are few losses. The engagement, if there is one, is an affair of a second. The proof is that in this action of cavalry against cavalry, the conquered alone loses men, and he loses generally few. The battle against infantry is alone the really deadly struggle. Like numbers of little chasseurs have routed heavy cuirassiers. How could they have done so if the others had not given way before their determination? The essential factor was, and always is, determination.

The cavalry's casualties are always much less than those of the infantry both from fire and from disease. Is it because the cavalry is the aristocratic arm? This explains why in long wars it improves much more than the infantry.

As there are few losses between cavalry and cavalry, so there is little fighting.

Hannibal's Numidians, like the Russian Cossacks, inspired a veritable terror by the incessant alarms they caused. They tired out without fighting and killed by surprise.

Why is the cavalry handled so badly?—It is true that infantry is not used better.—Because its rôle is one of movement, of morale, of morale and movement so united, that movement alone, often without a charge or shock action of any sort can drive the enemy into retreat, and, if followed closely, into rout. That is a result of the quickness of cavalry. One who knows how to make use of this quickness alone can obtain such results.

All writers on cavalry will tell you that the charge pushed home of two cavalry bodies and the shock at top speed do not exist. Always before the encounter, the weaker runs away, if there is not a face to face check. What becomes then of the MV squared? If this famous MV squared is an empty word, why then crush your horses under giants, forgetting that in the formula besides M there is V squared. In a charge, there is M, there is V squared, there is this and that. There is resolution, and I believe, nothing else that counts!

Cohesion and unity give force to the charge. Alignment is impossible at a fast gait where the most rapid pass the others. Only when the moral effect has been produced should the gait be increased to take advantage of it by falling upon an enemy already in disorder, in the act of fleeing. The cuirassiers charge at a trot. This calm steadiness frightens the enemy into an about face. Then they charge at his back, at a gallop.

They say that at Eckmühl, for every French cuirassier down, fourteen Austrians were struck in the back. Was it because they had no back-plate? It is evident that it was because they offered their backs to the blows.

Jomini speaks of charges at a trot against cavalry at a gallop. He cites Lasalle who used the trot and who, seeing cavalry approach at a gallop, would say: "There are lost men." Jomini insists on the effect of shock. The trot permits that compactness which the gallop breaks up. That may be true. But the effect is moral above all. A troop at the gallop sees a massed squadron coming towards it at a trot. It is surprised at first at such coolness. The material impulse of the gallop is superior; but there are no intervals, no gaps through which to penetrate the line in order to avoid the shock, the shock that overcomes men and horses. These men must be very resolute, as their close ranks do not permit them to escape by about facing. If they move at such a steady gait, it is because their resolution is also firm and they do not feel the need of running away, of diverting themselves by the unchecked speed of the unrestrained gallop, etc. [43]

Galloping men do not reason these things out, but they know them instinctively. They understand that they have before them a moral impulse superior to theirs. They become uneasy, hesitate. Their hands instinctively turn their horses aside. There is no longer freedom in the attack at a gallop. Some go on to the end, but three-fourths have already tried to avoid the shock. There is complete disorder, demoralization, flight. Then begins the pursuit at a gallop by the men who attacked at the trot.

The charge at a trot exacts of leaders and men complete confidence and steadfastness. It is the experience of battle only that can give this temper to all. But this charge, depending on a moral effect, will not always succeed. It is a question of surprise. Xenophon [44] recommended, in his work on cavalry operations, the use of surprise, the use of the gallop when the trot is customary, and vice-versa. "Because," he says, "agreeable or terrible, the less a thing is foreseen, the more pleasure or fright does it cause. This is nowhere seen better than in war, where every surprise strikes terror even to the strongest."

As a general rule, the gallop is and should be necessary in the charge; it is the winning, intoxicating gait, for men and horses. It is taken up at such a distance as may be necessary to insure its success, whatever it may cost in men and horses. The regulations are correct in prescribing that the charge be started close up. If the troopers waited until the charge was ordered, they would always succeed. I say that strong men, moved by pride or fear, by taking up too soon the charge against a firm enemy, have caused more charges to fail than to succeed. Keeping men in hand until the command "charge," seizing the precise instant for this command, are both difficult. They exact of the energetic leader domination over his men and a keen eye, at a moment when three out of four men no longer see anything, so that good cavalry leaders, squadron leaders in general are very rare. Real charges are just as rare.

Actual shock no longer exists. The moral impulse of one of the adversaries nearly always upsets the other, perhaps far off, perhaps a little nearer. Were this "a little nearer," face to face, one of the two troops would be already defeated before the first saber cut and would disentangle itself for flight. With actual shock, all would be thrown into confusion. A real charge on the one part or the other would cause mutual extermination. In practice the victor scarcely loses any one.

Observation demonstrates that cavalry does not close with cavalry; its deadly combats are those against infantry alone.

Even if a cavalryman waits without flinching, his horse will wish to escape, to shrink before the collision. If man anticipates, so does the horse. Why did Frederick like to see his center closed in for the assault? As the best guarantee against the instincts of man and horse.

The cavalry of Frederick had ordinarily only insignificant losses: a result of determination.

The men want to be distracted from the advancing danger by movement. The cavalrymen who go at the enemy, if left to themselves, would start at a gallop, for fear of not arriving, or of arriving exhausted and material for carnage. The same is true of the Arabs. Note what happened in 1864 to the cavalry of General Martineau. The rapid move relieves anxiety. It is natural to wish to lessen it. But the leaders are there, whom experience, whom regulations order to go slowly, then to accelerate progressively, so as to arrive with the maximum of speed. The procedure should be the walk, then the trot, after that the gallop, then the charge. But it takes a trained eye to estimate distance and the character of the terrain, and, if the enemy approaches, to pick the point where one should meet him. The nearer one approaches, the greater among the troops is the question of morale. The necessity of arriving at the greatest speed is not alone a mechanical question, since indeed one never clashes, it is a moral necessity. It is necessary to seize the moment at which the uneasiness of one's men requires the intoxication of the headlong charging gallop. An instant too late, and a too great anxiety has taken the upper hand and caused the hands of the riders to act on the horses; the start is not free; a number hide by remaining behind. An instant too soon: before arrival the speed has slowed down; the animation, the intoxication of the run, fleeting things, are exhausted. Anxiety takes the upper hand again, the hands act instinctively, and even if the start were unhampered, the arrival is not.

Frederick and Seidlitz were content when they saw the center of the charging squadron three and four ranks deep. It was as if they understood that with this compact center, as the first lines could not escape to the right or left, they were forced to continue straight ahead.

In order to rush like battering-rams, even against infantry, men and horses ought to be watered and fresh (Ponsomby's cavalry at Waterloo). If there is ever contact between cavalry, the shock is so weakened by the hands of the men, the rearing of the horses, the swinging of heads, that both sides come to a halt.

Only the necessity for carrying along the man and the horse at the supreme moment, for distracting them, necessitates the full gallop before attacking the enemy, before having put him to flight.

Charges at the gallop of three or four kilometers, suppose horses of bronze.

Because morale is not studied and because historical accounts are taken too literally, each epoch complains that cavalry forces are no longer seen charging and fighting with the sword, that too much prudence dictates running away instead of clashing with the enemy.

These plaints have been made ever since the Empire, both by the allies, and by us. But this has always been true. Man was never invulnerable. The charging gait has almost always been the trot. Man does not change. Even the combats of cavalry against cavalry today are deadlier than they were in the lamented days of chivalry.

The retreat of the infantry is always more difficult than that of the cavalry; the latter is simple. A cavalry repulsed and coming back in disorder is a foreseen, an ordinary happening; it is going to rally at a distance. It often reappears with advantage. One can almost say, in view of experience, that such is its rôle. An infantry that is repelled, especially if the action has been a hot one and the cavalry rushes in, is often disorganized for the rest of the day.

Even authors who tell you that two squadrons never collide, tell you continually: "The force of cavalry is in the shock." In the terror of the shock, Yes. In the shock, No! It lies only in determination. It is a mental and not a mechanical condition.

Never give officers and men of the cavalry mathematical demonstrations of the charge. They are good only to shake confidence. Mathematical reasoning shows a mutual collapse that never takes place. Show them the truth. Lasalle with his always victorious charge at a trot guarded against similar reasonings, which might have demonstrated to him mathematically that a charge of cuirassiers at a trot ought to be routed by a charge of hussars at a gallop. He simply told them: "Go resolutely and be sure that you will never find a daredevil determined enough to come to grips with you." It is necessary to be a daredevil in order to go to the end. The Frenchman is one above all. Because he is a good trooper in battle, when his commanders themselves are daredevils he is the best in Europe. (Note the days of the Empire, the remarks of Wellington, a good judge). If moreover, his leaders use a little head work, that never harms anything. The formula of the cavalry is R (Resolution) and R, and always R, and R is greater than all the MV squared in the world.

There is this important element in the pursuit of cavalry by cavalry. The pursued cannot halt without delivering himself up to the pursuer. The pursuer can always see the pursued. If the latter halts and starts to face about the pursuer can fall upon him before he is faced, and take him by surprise. But the pursued does not know how many are pursuing him. If he alone halts two pursuers may rush on him, for they see ahead of them and they naturally attack whoever tries to face about. For with the about face danger again confronts them. The pursuit is often instigated by the fear that the enemy will turn. The material fact that once in flight all together cannot turn again without risking being surprised and overthrown, makes the flight continuous. Even the bravest flee, until sufficient distance between them and the enemy, or some other circumstances such as cover or supporting troops, permits of a rally and a return to the offensive. In this case the pursuit may turn into flight in its turn.

Cavalry is insistent on attacking on an equal front. Because, if with a broader front, the enemy gives way before it, his wings may attack it and make it the pursued instead of the pursuer. The moral effect of resolution is so great that cavalry, breaking and pursuing a more numerous cavalry, is never pursued by the enemy wings. However the idea that one may be taken in rear by forces whom one has left on the flanks in a position to do so, has such an effect that the resolution necessary for an attack under these circumstances is rare.

Why is it that Colonel A—— does not want a depth formation for cavalry, he who believes in pressure of the rear ranks on the first? It is because at heart he is convinced that only the first rank can act in a cavalry charge, and that this rank can receive no impression, no speeding up, from those behind it.

There is debate as to the advantage of one or two ranks for the cavalry. This again is a matter of morale. Leave liberty of choice, and under varying conditions of confidence and morale one or the other will be adopted. There are enough officers for either formation.

It is characteristic of cavalry to advance further than infantry and consequently it exposes its flanks more. It then needs more reserves to cover its flanks and rear than does infantry. It needs reserves to protect and to support the pursuers who are almost always pursued when they return. With cavalry even more than infantry victory belongs to the last reserves held intact. The one with the reserves is always the one who can take the offensive. Tie to that, and no one can stand before you.

With room to maneuver cavalry rallies quickly. In deep columns it cannot.

The engagement of cavalry lasts only a moment. It must be reformed immediately. With a roll call at each reforming, it gets out of hand less than the infantry, which, once engaged, has little respite. There should be a roll call for cavalry, and for infantry after an advance, at each lull. There should be roll calls at drill and in field maneuvers, not that they are necessary but in order to become habituated to them. Then the roll call will not be forgotten on the day of action, when very few think of what ought to be done.

In the confusion and speed of cavalry action, man escapes more easily from surveillance. In our battles his action is increasingly individual and rapid. The cavalryman should not be left too free; that would be dangerous. Frequently in action troops should be reformed and the roll called. It would be an error not to do so. There might be ten to twenty roll calls in a day. The officers, the soldiers, would then have a chance to demand an accounting from each man, and might demand it the next day.

Once in action, and that action lasts, the infantryman of today escapes from the control of his officers. This is due to the disorder inherent in battle, to deployment, to the absence of roll calls, which cannot be held in action. Control, then, can only be in the hands of his comrades. Of modern arms infantry is the one in which there is the greatest need for cohesion.

Cavalry always fights very poorly and very little. This has been true from antiquity, when the cavalryman was of a superior caste to the infantryman, and ought to have been braver.

Anybody advancing, cavalry or infantry, ought to scout and reconnoiter as soon as possible the terrain on which it acts. Condé forgot this at Neerwinden. The 55th forgot it at Solferino. [45] Everybody forgets it. And from the failure to use skirmishers and scouts, come mistakes and disasters.

The cavalry has a rifle for exceptional use. Look out that this exception does not become the rule. Such a tendency has been seen. At the battle of Sicka, the first clash was marred by the lack of dash on the part of a regiment of Chasseurs d'Afrique, which after being sent off at the gallop, halted to shoot. At the second clash General Bugeaud charged at their head to show them how to charge.

A young Colonel of light cavalry, asked carbines for his cavalry. "Why? So that if I want to reconnoiter a village I can sound it from a distance of seven or eight hundred meters without losing anybody." What can you say to a man advancing such ideas? Certainly the carbine makes everybody lose common sense.

The work of light cavalry makes it inevitable that they be captured sometimes. It is impossible to get news of the enemy without approaching him. If one man escapes in a patrol, that is enough. If no one comes back, even that fact is instructive. The cavalry is a priceless object that no leader wants to break. However it is only by breaking it that results can be obtained.

Some authors think of using cavalry as skirmishers, mounted or dismounted. I suppose they advance holding the horse by the bridle? This appears to be to be an absurdity. If the cavalryman fires he will not charge. The African incident cited proves that. It would be better to give the cavalryman two pistols than a carbine.

The Americans in their vast country where there is unlimited room, used cavalry wisely in sending it off on distant forays to cut communications, make levies, etc. What their cavalry did as an arm in battle is unknown. The cavalry raids in the American war were part of a war directed against wealth, against public works, against resources. It was war of destruction of riches, not of men. The raiding cavalry had few losses, and inflicted few losses. The cavalry is always the aristocratic arm which loses very lightly, even if it risks all. At least it has the air of risking all, which is something at any rate. It has to have daring and daring is not so common. But the merest infantry engagements in equal numbers costs more than the most brilliant cavalry raid.

3. Cavalry Against Infantry

Cavalry knows how to fight cavalry. But how it fights infantry not one cavalry officer in a thousand knows. Perhaps not one of them knows. Go to it then gaily, with general uncertainty!

A military man, a participant in our great wars, recommends as infallible against infantry in line the charge from the flank, horse following horse. He would have cavalry coming up on the enemy's left, pass along his front and change direction so as to use its arms to the right. This cavalryman is right. Such charges should give excellent results, the only deadly results. The cavalryman can only strike to his right, and in this way each one strikes. Against ancient infantry such charges would have been as valuable as against modern infantry. This officer saw with his own eyes excellent examples of this attack in the wars of the Empire. I do not doubt either the facts he cites or the deductions he makes. But for such charges there must be officers who inspire absolute confidence in their men and dependable and experienced soldiers. There is necessary, in short, an excellent cavalry, seasoned by long wars, and officers and men of very firm resolution. So it is not astonishing that examples of this mode of action are rare. They always will be. They always require a head for the charge, an isolated head, and when he is actually about to strike, he will fall back into the formation. It seems to him that lost in the mass he risks less than when alone. Everybody is willing to charge, but only if all charge together. It is a case of belling the cat.

The attack in column on infantry has a greater moral action than the charge in line. If the first and second squadrons are repulsed, but the infantry sees a third charging through the dust, it will say "When is this going to stop?" And it will be shaken.

An extract from Folard: "Only a capable officer is needed to get the best results from a cavalry which has confidence in its movement, which is known to be good and vigorous, and also is equipped with excellent weapons. Such cavalry will break the strongest battalions, if its leader has sense enough to know its power and courage enough to use this power."

Breaking is not enough, and is a feat that costs more than it is worth if the whole battalion is not killed or taken prisoner, or at least if the cavalry is not immediately followed by other troops, charged with this task.

At Waterloo our cavalry was exhausted fruitlessly, because it acted without artillery or infantry support.

At Krasno, August 14, 1812, Murat, at the head of his cavalry could not break an isolated body of ten thousand Russian infantry which continually held him off by its fire, and retired tranquilly across the plain.

The 72nd was upset by cavalry at Solferino.

From ancient days the lone infantryman has always had the advantage over the lone cavalryman. There is no shadow of a doubt about this in ancient narrations. The cavalryman only fought the cavalryman. He threatened, harassed, troubled the infantryman in the rear, but he did not fight him. He slaughtered him when put to flight by other infantry, or at least he scattered him and the light infantry slaughtered him.

Cavalry is a terrible weapon in the hands of one who knows how to use it. Who can say that Epaminondas could have defeated the Spartans twice without his Thessalonian cavalry.

Eventually rifle and artillery fire deafen the soldier; fatigue overpowers him; he becomes inert; he hears commands no longer. If cavalry unexpectedly appears, he is lost. Cavalry conquers merely by its appearance. (Bismarck or Decker).

Modern cavalry, like ancient cavalry, has a real effect only on troops already broken, on infantry engaged with infantry, on cavalry disorganized by artillery fire or by a frontal demonstration. But against such troops its action is decisive. In such cases its action is certain and gives enormous results. You might fight all day and lose ten thousand men, the enemy might lose as many, but if your cavalry pursues him, it will take thirty thousand prisoners. Its role is less knightly than its reputation and appearance, less so than the rôle of infantry. It always loses much less than infantry. Its greatest effect is the effect of surprise, and it is thereby that it gets such astonishing results.

What formation should infantry, armed with modern weapons, take to guard against flank attacks by cavalry? If one fires four times as fast, if the fire is better sustained, one needs only a quarter as many men to guard a point against cavalry. Protection might be secured by using small groups, placed the range of a rifle shot apart and flanking each other, left on the flank of the advance. But they must be dependable troops, who will not be worried by what goes on behind them.

4. Armor and Armament

An armored cavalry is clearly required for moral reasons.

Note this with reference to the influence of cuirassiers (armored cavalrymen) on morale. At the battle of Renty, in 1554, Tavannes, a marshal, had with him his company armored in steel. It was the first time that such armor had been seen. Supported by some hundreds of fugitives who had rallied, he threw himself at the head of his company, on a column of two thousand German cavalry who had just thrown both infantry and cavalry into disorder. He chose his time so well that he broke and carried away these two thousand Germans, who fell back and broke the twelve hundred light horsemen who were supporting them. There followed a general flight, and the battle was won.

General Renard says "The decadence of cavalry caused the disappearance of their square formations in battle, which were characteristic in the seventeenth century." It was not the decadence of the cavalry but the abandonment of the cuirass and the perfecting of the infantry weapon to give more rapid fire. When cuirassiers break through they serve as examples, and emulation extends to others, who another time try to break through as they did.

Why cuirassiers? Because they alone, in all history, have charged and do charge to the end.

To charge to the end the cuirassiers need only half the courage of the dragoons, as their armor raises their morale one half. But since the cuirassiers have as much natural courage as the dragoons, for they are all the same men, it is proper to count the more on their action. Shall we have only one kind of cavalry? Which? If all our cavalry could wear the cuirass and at the same time do the fatiguing work of light cavalry, if all our horses could in addition carry the cuirass through such work, I say that there should be only cuirassiers. But I do not understand why the morale given by the cuirass should be lightly done away with, merely to have one cavalry without the cuirass.

A cavalryman armored completely and his horse partially, can charge only at a trot.

On the appearance of fire arms, cavalry, according to General Ambert, an author of the past, covered itself with masses of armor resembling anvils rather than with cuirasses. It was at that time the essential arm. Later as infantry progressed the tactics changed, it needed more mobility. Permanent armies began to be organized by the State. The State thought less of the skin of the individual than of economy and mobility and almost did away with cuirassiers. The cuirass has always given, and today more than ever it will give, confidence to the cavalryman. Courage, dash, and speed have a value beyond that of mere mass. I leave aside mathematical discussions which seem to me to have nothing in common with battle conditions. I would pick to wear the cuirass the best men in the army, big chested, red-blooded, strong limbed, the foot chasseurs. I would organize a regiment of light cuirassiers for each of our divisions. Men and horses, such a cavalry would be much more robust and active than our present cuirassiers. If our armored cavalry is worth more than any other arm by its dash in battle, this cavalry would be worth twice as much. But how would these men of small stature get into the saddle? To this serious objection I answer, "They will arrange it." And this objection, which I do not admit, is the only one that can be made against the organization of a light armored cavalry, an organization that is made imperative by the improvement in weapons. The remainder of those chasseur battalions which furnish cuirassiers, should return to the infantry, which has long demanded them, and hussars and dragoons, dismounted in the necessary number will also be welcomed by the infantry.

As for the thrust, the thrust is deadlier than the cut. You do not have to worry about lifting your arm; you thrust. But it is necessary that the cavalryman be convinced that to parry a vertical cut is folly. This can be done by his officers, by those who have had experience, if there are any such in peace times. This is not easy. But in this respect, as in all others, the advantage lies with the brave. A cavalry charge is a matter of morale above all. It is identical in its methods, its effects, with the infantry charge. All the conditions to be fulfilled in the charge (walk, trot, gallop, charge, etc.) have a reason bearing on morale. These reasons have already been touched on.

Roman discipline and character demand tenacity. The hardening of the men to fatigue, and a good organization, giving mutual support, produced that tenacity, against which the bravest could not stand. The exhausting method of powerful strokes used by the Gauls could not last long against the skillful, terrible and less fatiguing method of fighting by the thrust.

The Sikh cavalrymen of M. Nolan armed with dragoon sabers sharpened by themselves, liked the cut. They knew nothing about methods of swordsmanship; they did not practice. They said "A good saber and a willingness to use it are enough." True, True!

There is always discussion as to the lance or the saber. The lance requires skillful vigorous cavalrymen, good horsemen, very well drilled, very adroit, for the use of the lance is more difficult than that of the straight sword, especially if the sword is not too heavy. Is not this an answer to the question? No matter what is done, no matter what methods are adopted, it must always be remembered that our recruits in war time are sent into squadrons as into battalions, with a hasty and incomplete training. If you give them lances, most of them will just have sticks in their hands, while a straight sword at the end of a strong arm is at the same time simple and terrible. A short trident spear, with three short points just long enough to kill but not only enough to go through the body, would remain in the body of the man and carry him along. It would recoil on the cavalryman who delivered the blow, he would be upset by the blow himself. But the dragoon must be supported by the saddle, and as he had kept hold of the shaft he would be able to disengage the fork which had pierced the body some six inches. No cavalry of equal morale could stand against a cavalry armed with such forked spears.

As between forks and lances, the fork would replace the lance. That is, of course, for beginners in mounted fencing. But the fork! It would be ridiculous, not military!

With the lance one always figures without the horse, whose slightest movement diverts the lance so much. The lance is a weapon frightful even to the mounted man who uses it properly. If he sticks an enemy at the gallop, he is dismounted, torn off by the arm attached to the lance which remains in the body of his enemy.

Cavalry officers and others who seek examples in "Victories and Conquests," in official reports, in "Bazancourt" are too naïve. It is hard to get at the truth. In war, in all things, we take the last example which we have witnessed. And now we want lances, which we do not know how to use, which frighten the cavalryman himself and pluck him from the saddle if he sticks anybody. We want no more cuirasses; we want this and that. We forget that the last example gives only a restricted number of instances relating to the matter in question.

It appears, according to Xenophon, that it was not easy to throw the dart from horseback. He constantly recommends obtaining as many men as possible who know how to throw the dart. He recommends leaning well back to avoid falling from the horse in the charge. In reading Xenophon it is evident that there was much falling from the horse.

It appears that in battle there is as great difficulty in handling the saber as in handling the bayonet. Another difficulty for the cavalryman lies in the handling of the musket. This is seen in the handling of the regulation weapon of the Spahis. There is only one important thing for the cavalryman, to be well seated. Men should be on horseback for hours at a time, every day, from their arrival in the organization. If the selection of those who know something about horses was not neglected in the draft, and if such men were, made cavalrymen, the practical training of the greater number would be much more rapidly concluded. I do not speak of the routine of the stable. Between mounted drills, foot drills might be gone through with in a snappy, free fashion, without rigidity, with daily increasing speed. Such drills would instruct cavalrymen more rapidly than the restricted method employed.

A dragoon horse carries in campaign with one day's food three hundred and eight pounds, without food or forage two hundred and seventy seven pounds. How can such horses carry this and have speed?

Seek the end always, not the means! Make a quarter of your cavalrymen into muleteers, a quarter of your horses into pack animals. You will thus secure, for the remaining three quarters unquestioned vigor. But how will you make up these pack trains? You will have plenty of wounded horses after a week of campaign.

CHAPTER IV

ARTILLERY

If artillery did not have a greater range than the rifle, we could not risk separating it far from its support, as it would have to wait until the enemy was but four or five hundred paces away to fire on him. But the more its range is increased, the further away it can be placed from its support.

The greater the range of artillery, the greater freedom of action from the different arms, which no longer have to be side by side to give mutual support.

The greater the range of artillery, the easier it is to concentrate its fire. Two batteries fifteen hundred meters apart can concentrate on a point twelve hundred meters in front of and between them. Before the range was so long they had to be close together, and the terrain did not always lend itself to this.

Furthermore, do not support a piece by placing infantry just behind or alongside of it, as is done three-quarters of the time at maneuvers. On the contrary hide the infantry to the right or left and far behind, cover it without worrying too much about distance and let the artillery call for help if they think that the piece is in danger of being lost. Why should infantry be placed too close, and consequently have its advance demoralized? This will throw away the greatest advantage that we Frenchmen have in defense, that of defending ourselves by advancing, with morale unimpaired, because we have not suffered heavy losses at a halt. There is always time to run to the defense of artillery. To increase the moral effect advance your supports in formation. Skirmishers can also be swiftly scattered among the batteries. These skirmishers, in the midst of the guns will not have to fear cavalry. Even if they are assailed by infantry it will not be such a terrible thing. The engagement will merely be one between skirmishers, and they will be able to take cover behind the pieces, firing against the enemy who is coming up in the open.

Guibert, I believe, held that artillery should not worry whether it was supported or not; that it should fire up to the last minute, and finally abandon the pieces, which supporting troops might or might not recapture. These supporting troops should not be too close. It is easier to defend pieces, to take them back even, by advancing on an enemy dispersed among them, than to defend them by standing fast after having participated in the losses suffered by the artillery under fire. (Note the English in Spain. The system of having artillery followed by infantry platoons is absurd.)

Artillery in battle has its men grouped around the pieces, stationary assembly points, broadly distributed, each one having its commander and its cannoneers, who are always the same. Thus there is in effect a roll call each time artillery is put into battery. Artillery carries its men with it; they cannot be lost nor can they hide. If the officer is brave, his men rarely desert him. Certainly, in all armies, it is in the artillery that the soldier can best perform his duty.

As General Leboeuf tells us, four batteries of artillery can be maneuvered, not more. That is all right. Here is the thing in a nut-shell. Four battalions is a big enough command for a colonel. A general has eight battalions. He gets orders, "General, do so and so." He orders, "Colonel, do so and so." So that without any maneuvers being laid down for more than four battalions, as many battalions as you like can be maneuvered and drilled.

CHAPTER V

COMMAND, GENERAL STAFF, AND ADMINISTRATION

There are plenty of carefree generals, who are never worried nor harassed. They do not bother about anything. They say, "I advance. Follow me." The result is an incredible disorder in the advance of columns. If ten raiders should fall on the column with a shout, this disorder would become a rout, a disaster. But these gentlemen never bother with such an eventuality. They are the great men of the day, until the moment that some disaster overwhelms them.

Cavalry is no more difficult to work with than infantry. According to some military authors, a cavalry general ought to have the wisdom of the phoenix. The perfect one should have. So should the perfect infantry general. Man on horseback and man afoot is always the same man. Only, the infantry general rarely has to account for the losses in his command, which may have been due to faulty or improper handling. The cavalry general does have to do this. (We shall lay aside the reasons why.) The infantry general has six chances for real battle to one for the cavalry general. These are the two reasons why, from the beginning of a war, more initiative is found in infantry than in cavalry generals. General Bugeaud might have made a better cavalry general than an infantry general. Why? Because he had immediate decision and firm resolution. There is more need for resolution in the infantryman than in the cavalryman. Why? There are many reasons, which are matters of opinion.

In short, the infantryman is always more tired than the cavalryman. His morale is therefore harder to keep up. I believe therefore that a good infantry general is rarer than one of cavalry. Also, the resolution of an infantry general does not have to last for a moment only; it has to endure for a long, long time.

Good artillery generals are common. They are less concerned with morale than with other things, such as material results. They have less need to bother about the morale of their troops, as combat discipline is always better with them than with the other arms. This is shown elsewhere.

Brigadier generals ought to be in their prescribed places. Very well, but the most of them are not and never have been. They were required to be in place at the battle of Moscow, but, as they were so ordered there, it is evident that they were not habitually in place. They are men; and their rank, it seems to them, ought to diminish rather than increase the risks they have to run. And, then, in actual engagement, where is their prescribed place?

When one occupies a high command there are many things which he does not see. The general-in-chief, even a division commander, can only escape this failing by great activity, moved by strict conscientiousness and aided by clairvoyance. This failing extends to those about him, to his heads of services. These men live well, sleep well; the same must be true of all! They have picked, well-conditioned horses; the roads are excellent! They are never sick; the doctors must be exaggerating sickness! They have attendants and doctors; everybody must be well looked after! Something happens which shows abominable negligence, common enough in war. With a good heart and a full belly they say, "But this is infamous, unheard of! It could not have happened! It is impossible! etc."

To-day there is a tendency, whose cause should be sought, on the part of superiors to infringe on the authority of inferiors. This is general. It goes very high and is furthered by the mania for command, inherent in the French character. It results in lessening the authority of subordinate officers in the minds of their soldiers. This is a grave matter, as only the firm authority and prestige of subordinate officers can maintain discipline. The tendency is to oppress subordinates; to want to impose on them, in all things, the views of the superior; not to admit of honest mistakes, and to reprove them as faults; to make everybody, even down to the private, feel that there is only one infallible authority. A colonel, for instance, sets himself up as the sole authority with judgment and intelligence. He thus takes all initiative from subordinate officers, and reduces them to a state of inertia, coming from their lack of confidence in themselves and from fear of being severely reproved. How many generals, before a regiment, think only of showing how much they know! They lessen the authority of the colonel. That is nothing to them. They have asserted their superiority, true or false; that is the essential. With cheeks puffed out, they leave, proud of having attacked discipline.

This firm hand which directs so many things is absent for a moment. All subordinate officers up to this moment have been held with too strong a hand, which has kept them in a position not natural to them. Immediately they are like a horse, always kept on a tight rein, whose rein is loosened or missing. They cannot in an instant recover that confidence in themselves, that has been painstakingly taken away from them without their wishing it. Thus, in such a moment conditions become unsatisfactory, the soldier very quickly feels that the hand that holds him vacillates.

"Ask much, in order to obtain a little," is a false saying, a source of errors, an attack on discipline. One ought to obtain what one asks. It is only necessary to be moderately reasonable and practical.

In following out this matter, one is astonished at the lack of foresight found in three out of four officers. Why? Is there anything so difficult about looking forward a little? Are three-quarters of the officers so stupid? No! It is because their egoism, generally frankly acknowledged, allow them to think only of who is looking at them. They think of their troops by chance perhaps, or because they have to. Their troops are never their preoccupation, consequently they do not think about them at all. A major in command of an organization in Mexico, on his first march in a hot country, started without full canteens, perhaps without canteens at all, without any provision for water, as he might march in France. No officer in his battalion called his attention to the omission, nor was more foresighted than he. In this first march, by an entire lack of foresight in everything, he lost, in dead, half of his command. Was he reduced? No! He was made a lieutenant-colonel.

Officers of the general staff learn to order, not to command. "Sir, I order," a popular phrase, applies to them.

The misfortune is not that there is a general staff, but that it has achieved command. For it always has commanded, in the name of its commanders it is true, and never obeyed, which is its duty. It commands in fact. So be it! But just the same it is not supposed to.

Is it the good quality of staffs or that of combatants that makes the strength of armies? If you want good fighting men, do everything to excite their ambition, to spare them, so that people of intelligence and with a future will not despise the line but will elect to serve in it. It is the line that gives you your high command, the line only, and very rarely the staff. The staff, however, dies infrequently, which is something. Do they say that military science can only be learned in the general staff schools? If you really want to learn to do your work, go to the line.

To-day, nobody knows anything unless he knows how to argue and chatter. A peasant knows nothing, he is a being unskilled even in cultivating the soil. But the agriculturist of the office is a farmer emeritus, etc. Is it then believed that there is ability only in the general staff? There is the assurance of the scholar there, of the pedagogue who has never practiced what he preaches. There is book learning, false learning when it treats of military matters. But knowledge of the real trade of a soldier, knowledge of what is possible, knowledge of blows given and received, all these are conspicuously absent.

Slowness of promotion in the general staff as compared to its rapidity in the line might make many men of intelligence, of head and heart, pass the general staff by and enter the line to make their own way. To be in the line would not then be a brevet of imbecility. But to-day when general staff officers rank the best of the line, the latter are discouraged and rather than submit to this situation, all who feel themselves fitted for advancement want to be on the general staff. So much the better? So much the worse. Selection is only warranted by battle.

How administrative deceits, in politics or elsewhere, falsify the conclusions drawn from a fact!

In the Crimea one hundred per cent. of the French operated upon succumbed, while only twenty-seven per cent. of the English operated upon died. That was attributed to the difference in temperament! The great cause of this discrepancy was the difference in care. Our newspapers followed the self-satisfied and rosy statements given out by our own supply department. They pictured our sick in the Crimea lying in beds and cared for by sisters of charity. The fact is that our soldiers never had sheets, nor mattresses, nor the necessary changes of clothes in the hospitals; that half, three-quarters, lay on mouldy straw, on the ground, under canvass. The fact is, that such were the conditions under which typhus claimed twenty-five to thirty thousand of our sick after the siege; that thousands of pieces of hospital equipment were offered by the English to our Quartermaster General, and that he refused them! Everybody ought to have known that he would! To accept such equipment was to acknowledge that he did not have it. And he ought to have had it. Indeed he did according to the newspapers and the Quartermaster reports. There were twenty-five beds per hospital so that it could be said, "We have beds!" Each hospital had at this time five hundred or more sick.

These people are annoyed if they are called hypocrites. While our soldiers were in hospitals, without anything, so to speak, the English had big, well-ventilated tents, cots, sheets, even night stands with urinals. And our men had not even a cup to drink from! Sick men were cared for in the English hospitals. They might have been in ours, before they died, which they almost always did.

It is true that we had the typhus and the English had not. That was because our men in tents had the same care as in our hospitals, and the English the same care as in their hospitals.

Read the war reports of supply departments and then go unexpectedly to verify them in the hospitals and storehouses. Have them verified by calling up and questioning the heads of departments, but question them conscientiously, without dictating the answers. In the Crimea, in May of the first year, we were no better off than the English who complained so much, Who has dared to say, however, that from the time they entered the hospital to the time that they left it, dead, evacuated, or cured, through fifteen or twenty days of cholera or typhus, our men lay on the same plank, in the same shoes, drawers, shirts and clothing that they brought in with them? They were in a state of living putrefaction that would by itself have killed well men! The newspapers chanted the praises of the admirable French administration. The second winter the English had no sick, a smaller percentage than in London. But to the eternal shame of the French command and administration we lost in peace time, twenty-five to thirty thousand of typhus and more than one thousand frozen to death. Nevertheless, it appeared that we had the most perfect administration in the world, and that our generals, no less than our administration, were full of devoted solicitude to provide all the needs of the soldier. That is an infamous lie, and is known as such, let us hope.

The Americans have given us a good example. The good citizens have gone themselves to see how their soldiers were treated and have provided for them themselves. When, in France, will good citizens lose faith in this best of administrations which is theirs? When will they, confident in themselves, do spontaneously, freely, what their administration cannot and never will be able to do?

The first thing disorganized in an army is the administration. The simplest foresight, the least signs even of order disappear in a retreat. (Note Russia-Vilna).

In the Crimea, and everywhere more or less, the doctor's visit was without benefit to the patient. It was made to keep up his spirits, but could not be followed by care, due to lack of personnel and material. After two or three hours of work, the doctor was exhausted.

In a sane country the field and permanent hospitals ought to be able to handle one-fifth of the strength at least. The hospital personnel of to-day should be doubled. It is quickly cut down, and it ought to have time, not only to visit the sick, but to care for them, feed them, dose and dress them, etc.

CHAPTER VI

SOCIAL AND MILITARY INSTITUTIONS.

NATIONAL CHARACTERISTICS.

Man's admiration for the great spectacles of nature is the admiration for force. In the mountains it is mass, a force, that impresses him, strikes him, makes him admire. In the calm sea it is the mysterious and terrible force that he divines, that he feels in that enormous liquid mass; in the angry sea, force again. In the wind, in the storm, in the vast depth of the sky, it is still force that he admires.

All these things astounded man when he was young. He has become old, and he knows them. Astonishment has turned to admiration, but always it is the feeling of a formidable force which compels his admiration. This explains his admiration for the warrior.

The warrior is the ideal of the primitive man, of the savage, of the barbarian. The more people rise in moral civilization, the lower this ideal falls. But with the masses everywhere the warrior still is and for a long time will be the height of their ideals. This is because man loves to admire the force and bravery that are his own attributes. When that force and bravery find other means to assert themselves, or at least when the crowd is shown that war does not furnish the best examples of them, that there are truer and more exalted examples, this ideal will give way to a higher one.

Nations have an equal sovereignty based on their existence as states. They recognize no superior jurisdiction and call on force to decide their differences. Force decides. Whether or not might was right, the weaker bows to necessity until a more successful effort can be made. (Prud'homme). It is easy to understand Gregory VII's ideas on the subject.

In peace, armies are playthings in the hands of princes. If the princes do not know anything about them, which is usually the case, they disorganize them. If they understand them, like the Prince of Prussia, they make their armies strong for war.

The King of Prussia and the Prussian nobility, threatened by democracy, have had to change the passion for equality in their people into a passion for domination over foreign nations. This is easily done, when domination is crowned with success, for man, who is merely the friend of equality is the lover of domination. So that he is easily made to take the shadow for the substance. They have succeeded. They are forced to continue with their system. Otherwise their status as useful members of society would be questioned and they would perish as leaders in war. Peace spells death to a nobility. Consequently nobles do not desire it, and stir up rivalries among peoples, rivalries which alone can justify their existence as leaders in war, and consequently as leaders in peace. This is why the military spirit is dead in France. The past does not live again. In the spiritual as in the physical world, what is dead is dead. Death comes only with the exhaustion of the elements, the conditions which are necessary for life. For these reasons revolutionary wars continued into the war with Prussia. For these reasons if we had been victorious we would have found against us the countries dominated by nobilities, Austria, Russia, England. But with us vanquished, democracy takes up her work in all European countries, protected in the security which victory always gives to victors. This work is slower but surer than the rapid work of war, which, exalting rivalries, halts for a moment the work of democracy within the nations themselves. Democracy then takes up her work with less chance of being deterred by rivalry against us. Thus we are closer to the triumph of democracy than if we had been victors. French democracy rightfully desires to live, and she does not desire to do so at the expense of a sacrifice of national pride. Then, since she will still be surrounded for a long time by societies dominated by the military element, by the nobility, she must have a dependable army. And, as the military spirit is on the wane in France, it must be replaced by having noncommissioned officers and officers well paid. Good pay establishes position in a democracy, and to-day none turn to the army, because it is too poorly paid. Let us have well paid mercenaries. By giving good pay, good material can be secured, thanks to the old warrior strain in the race. This is the price that must be paid for security.

The soldier of our day is a merchant. So much of my flesh, of my blood, is worth so much. So much of my time, of my affections, etc. It is a noble trade, however, perhaps because man's blood is noble merchandise, the finest that can be dealt in.

M. Guizot says "Get rich!" That may seem cynical to prudes, but it is truly said. Those who deny the sentiment, and talk to-day so loftily, what do they advise? If not by words, then by example they counsel the same thing; and example is more contagious. Is not private wealth, wealth in general, the avowed ambition sought by all, democrats and others? Let us be rich, that is to say, let us be slaves of the needs that wealth creates.

The Invalides in France, the institutions for pensioners, are superb exhibits of pomp and ostentation. I wish that their founding had been based on ideas of justice and Christianity and not purely on military-political considerations. But the results are disastrous to morality. This collection of weaklings is a school of depravity, where the invalided soldier loses in vice his right to respect.

Some officers want to transform regiments into permanent schools for officers of all ranks, with a two-hour course each day in law, military art, etc. There is little taste for military life in France; such a procedure would lessen it. The leisure of army life attracts three out of four officers, laziness, if you like. But such is the fact. If you make an officer a school-boy all his life he will send his profession to the devil, if he can. And those who are able to do so, will in general be those who have received the best education. An army is an extraordinary thing, but since it is necessary, there should be no astonishment that extraordinary means must be taken to keep it up; such as offering in peace time little work and a great deal of leisure. An officer is a sort of aristocrat, and in France we have no finer ideal of aristocratic life than one of leisure. This is not a proof of the highest ideals, nor of firmness of character. But what is to be done about it?

From the fact that military spirit is lacking in our nation (and officers are with greater difficulty than ever recruited in France) it does not follow that we shall not have to engage in war. Perhaps the contrary is true.

It is not patriotic to say that the military spirit is dead in France? The truth is always patriotic. The military spirit died with the French nobility, perished because it had to perish, because it was exhausted, at the end of its life. That only dies which has no longer the sap of life, and can no longer live. If a thing is merely sick it can return to health. But who can say that of the French nobility? An aristocracy, a nobility that dies, dies always by its own fault; because it no longer performs its duties; because it fails in its task; because its functions are of no more value to the state; because there is no longer any reason for its existence in a society, whose final tendency is to suppress its functions.

After 1789 had threatened our patriotism, the natural desire for self-protection revived the military spirit in the nation and in the army. The Empire developed this movement, changed the defensive military spirit to the offensive, and used it with increasing effect up to 1814 or 1815. The military spirit of the July Restoration was a reminiscence, a relic of the Empire, a form of opposition to government by liberalism instead of democracy. It was really the spirit of opposition and not the military spirit, which is essentially conservative.

There is no military spirit in a democratic society, where there is no aristocracy, no military nobility. A democratic society is antagonistic to the military spirit.

The military spirit was unknown to the Romans. They made no distinction between military and civil duties. I think that the military air dates from the time that the profession of arms became a private profession, from the time of the bravos, the Italian condottieri, who were more terrifying to civilians than to the enemy. When the Romans said "cedant arma togae," they did not refer to civil officials and soldiers; the civil officials were then soldiers in their turn; professional soldiers did not exist. They meant "might gives way to right."

Machiavelli quotes a proverb, "War makes thieves and peace has them hanged." The Spaniards in Mexico, which has been in rebellion for forty years, are more or less thieves. They want to continue to ply the trade. Civil authority exists no longer with them, and they would look on obedience to such an authority as shameful. It is easy to understand the difficulty of organizing a peaceful government in such a country. Half the population would have to hang the other half. The other half does not want to be hanged.

We are a democratic society; we become less and less military. The Prussian, Russian, Austrian aristocracies which alone make the military spirit of those states, feel in our democratic society an example which threatens their existence, as nobility, as aristocracy. They are our enemies and will be until they are wiped, out, until the Russian, Austrian and Prussian states become democratic societies, like ours. It is a matter of time.

The Prussian aristocracy is young. It has not been degenerated by wealth, luxury and servility of the court. The Prussian court is not a court in the luxurious sense of the word. There is the danger.

Meanwhile Machiavellian doctrines not being forbidden to aristocracies, these people appeal to German Jingoism, to German patriotism, to all the passions which move one people who are jealous of another. All this is meant to hide under a patriotic exterior their concern for their own existence as an aristocracy, as a nobility.

The real menace of the day is czarism, stronger than the czars themselves, which calls for a crusade to drive back Russia and the uncultured Slav race.

It is time that we understood the lack of power in mob armies; that we recall to mind the first armies of the revolution that were saved from instant destruction only by the lack of vigor and decision in European cabinets and armies. Look at the examples of revolutionaries of all times, who have all to gain and cannot hope for mercy. Since Spartacus, have they not always been defeated? An army is not really strong unless it is developed from a social institution. Spartacus and his men were certainly terrible individual fighters. They were gladiators used to struggle and death. They were prisoners, barbarian slaves enraged by their loss of liberty, or escaped serfs, all men who could not hope for mercy. What more terrible fighters could be imagined? But discipline, leadership, all was improvised and could not have the firm discipline coming down from the centuries and drawn from the social institutions of the Romans. They were conquered. Time, a long time, is needed to give to leaders the habit of command and confidence in their authority—to the soldiers confidence in their leaders and in their fellows. It is not enough to order discipline. The officers must have the will to enforce it, and its vigorous enforcement must instill subordination in the soldiers. It must make them fear it more than they fear the enemy's blows.

How did Montluc fight, in an aristocratic society? Montluc shows us, tells us. He advanced in the van of the assault, but in bad places he pushed in front of him a soldier whose skin was not worth as much as was his. He had not the slightest doubt or shame about doing this. The soldier did not protest, the propriety of the act was so well established. But you, officers, try that in a democratic army, such as we have commenced to have, such as we shall later have!

In danger the officer is no better than the soldier. The soldier is willing enough to advance, but behind his officer. Also, his comrades' skin is no more precious than is his, they must advance too. This very real concern about equality in danger, which seeks equality only, brings on hesitation and not resolution. Some fools may break their heads in closing in, but the remainder will fire from a distance. Not that this will cause fewer losses, far from it.

Italy will never have a really firm army. The Italians are too civilized, too fine, too democratic in a certain sense of the word. The Spaniards are the same. This may cause laughter, but it is true. The French are indeed worthy sons of their fathers, the Gauls. War, the most solemn act in the life of a nation, the gravest of acts, is a light thing to them. The good Frenchman lets himself be carried away, inflamed by the most ridiculous feats of arms into the wildest enthusiasm. Moreover he interprets the word "honor" in a fashion all his own. An expedition is commenced without sufficient reason, and good Frenchmen, who do not know why the thing is done, disapprove. But presently blood is spilled. Good sense and justice dictate that this spilled blood should taint those responsible for an unjust enterprise. But jingoism says "French blood has been spilled: Honor is at stake!" And millions of gold, which is the unit of labor, millions of men, are sacrificed to a ridiculous high-sounding phrase.

Whence comes this tendency toward war which characterizes above all the good citizen, the populace, who are not called upon personally to participate? The military man is not so easily swayed. Some hope for promotion or pension, but even they are sobered by their sense of duty. It comes from the romance that clothes war and battle, and that has with us ten times more than elsewhere, the power of exciting enthusiasm in the people. It would be a service to humanity and to one's people to dispell this illusion, and to show what battles are. They are buffooneries, and none the less buffooneries because they are made terrible by the spilling of blood. The actors, heroes in the eyes of the crowd, are only poor folk torn between fear, discipline and pride. They play some hours at a game of advance and retreat, without ever meeting, closing with, even seeing closely, the other poor folks, the enemy, who are as fearful as they but who are caught in the same web of circumstance.

What should be considered is how to organize an army in a country in which there is at the same time national and provincial feeling. Such a country is France, where there is no longer any necessity for uniting national and provincial feeling by mixing up the soldiers. In France, will the powerful motif of pride, which comes from the organization of units from particular provinces, be useful? From the fusion of varying elements comes the character of our troops, which is something to be considered. The make-up of the heavy cavalry should be noted. It has perhaps too many Germans and men from the northern provinces.

French sociability creates cohesion in French troops more quickly than could be secured in troops in other nations. Organization and discipline have the same purpose. With a proud people like the French, a rational organization aided by French sociability can often secure desired results without it being necessary to use the coercion of discipline.

Marshal de Gouvion-Saint Cyr said, "Experienced soldiers know and others ought to know that French soldiers once committed to the pursuit of the enemy will not return to their organization that day until forced back into it by the enemy. During this time they must be considered as lost to the rest of the army."

At the beginning of the Empire, officers, trained in the wars of the Revolution by incessant fighting, possessed great firmness. No one would wish to purchase such firmness again at the same price. But in our modern wars the victor often loses more than the vanquished, apart from the temporary loss in prisoners. The losses exceed the resources in good men, and discourage the exhausted, who appear to be very numerous, and those who are skilled in removing themselves from danger. Thus we fall into disorder. The Duke of Fezensac, testifying of other times, shows us the same thing that happens to-day. Also to-day we depend only on mass action, and at that game, despite the cleverest strategic handling, we must lose all, and do.

French officers lack firmness but have pride. In the face of danger they lack composure, they are disconcerted, breathless, hesitant, forgetful, unable to think of a way out. They call, "Forward, forward." This is one of the reasons why handling a formation in line is difficult, especially since the African campaigns where much is left to the soldier.

The formation in rank is then an ideal, unobtainable in modern war, but toward which we should strive. But we are getting further away from it. And then, when habit loses its hold, natural instinct resumes its empire. The remedy lies in an organization which will establish cohesion by the mutual acquaintanceship of all. This will make possible mutual surveillance, which has such power over French pride.

It might be said that there are two kinds of war, that in open country, and in the plain, and that of posts garrisoning positions in broken country. In a great war, with no one occupying positions, we should be lost immediately. Marshal Saxe knew us well when he said that the French were best for a war of position. He recognized the lack of stability in the ranks.

On getting within rifle range the rank formation tends to disappear. You hear officers who have been under fire say "When you get near the enemy, the men deploy as skirmishers despite you. The Russians group under fire. Their holding together is the huddling of sheep moved by fear of discipline and of danger." There are then two modes of conduct under fire, the French and the Russian.

The Gauls, seeing the firmness of the Roman formation, chained themselves together, making the first rank unbreakable and tying living to dead. This forbade the virtue they had not divined in the Roman formation, the replacement of wounded and exhausted by fresh men. From this replacement came the firmness which seemed so striking to the Gauls. The rank continually renewed itself.

Why does the Frenchman of to-day, in singular contrast to the Gaul, scatter under fire? His natural intelligence, his instinct under the pressure of danger causes him to deploy.

His method must be adopted. In view of the impossibility to-day of the Roman Draconian discipline which put the fear of death behind the soldier, we must adopt the soldier's method and try to put some order into it. How? By French discipline and an organization that permits of it.

Broken, covered country is adapted to our methods. The zouaves at Magenta could not have done so well on another kind of ground. [46]

Above all, with modern weapons, the terrain to be advanced over must be limited in depth.

How much better modern tactics fit the impatient French character! But also how necessary it is to guard against this impatience and to keep supports and reserves under control.

It should be noted that German or Gallic cavalry was always better than Roman cavalry, which could not hold against it, even though certainly better armed. Why was this? Because decision, impetuosity, even blind courage, have more chance with cavalry than with infantry. The defeated cavalry is the least brave cavalry. (A note for our cavalry here!) It was easier for the Gauls to have good cavalry than it is for us, as fire did not bother them in the charge.

The Frenchman has more qualities of the cavalryman than of the infantryman. Yet French infantry appears to be of greater value. Why? Because the use of cavalry on the battlefield requires rare decision and the seizing of the crucial opportunity. If the cavalryman has not been able to show his worth, it is the fault of his leaders. French infantry has always been defeated by English infantry. In cavalry combat the English cavalry has always fled before the French in those terrible cavalry battles that are always flights. Is this because in war man lasts longer in the cavalry and because our cavalrymen were older and more seasoned soldiers than our infantry? This does not apply to us only. If it is true for our cavalrymen, it is also true for the English cavalrymen. The reason is that on the field of battle the rôle of the infantryman against a firm adversary requires more coolness and nerve than does the rôle of the cavalryman. It requires the use of tactics based on an understanding of the national characteristics of ourselves and of our enemies. Against the English the confidence in the charge that is implanted in our brains, was completely betrayed. The rôle of cavalry against cavalry is simpler. The French confidence in the charge makes good fighting cavalry, and the Frenchman is better fitted than any other for this role. Our cavalry charge better than any other. That is the whole thing, on the battle field it is understood. As they move faster than infantry, their dash, which has its limits, is better preserved when they get up to the enemy.

The English have always fled before our cavalry. This proves that, strong enough to hold before the moral impulse of our infantry, they were not strong enough to hold before the stronger impulse of cavalry.

We ought to be much better cavalrymen than infantrymen, because the essential in a cavalryman is a fearless impetuosity. That is for the soldier. The cavalry leader ought to use this trait without hesitation, at the same time taking measures to support it and to guard against its failings. The attack is always, even on the defensive, an evidence of resolution, and gives a moral ascendancy. Its effect is more immediate with cavalry, because the movements of cavalry are more rapid and the moral effect has less time to be modified by reflection. To insure that the French cavalry be the best in Europe, and a really good cavalry, it needs but one thing, to conform to the national temperament, to dare, to dare, and to advance.

One of the singular features of French discipline is that on the road, especially in campaign the methods of punishment for derelictions become illusory, impractical. In 1859 there were twenty-five thousand skulkers in the Army in Italy. The soldier sees this immediately and lack of discipline ensues. If our customs do not permit of Draconian discipline, let us replace that moral coercion by another. Let us insure cohesion by the mutual acquaintanceship of men and officers; let us call French sociability to our aid.

With the Romans discipline was severest and most rigidly enforced in the presence of the enemy. It was enforced by the soldiers themselves. To-day, why should not the men in our companies watch discipline and punish themselves. They alone know each other, and the maintenance of discipline is so much to their interest as to encourage them to stop skulking. The twenty-five thousand men who skulked in Italy, all wear the Italian medal. They were discharged with certificates of good conduct. This certificate, in campaign should be awarded by the squad only. In place of that, discipline must be obtained somehow, and it is placed as an additional burden on the officer. He above all has to uphold it. He is treated without regard for his dignity. He is made to do the work of the non-commissioned officer. He is used as fancy dictates.

This cohesion which we hope for in units from squad to company, need not be feared in other armies. It cannot develop to the same point and by the same methods with them as with us. Their make-up is not ours, their character is different. This individuality of squads and companies comes from the make-up of our army and from French sociability.

Is it true that the rations of men and horses are actually insufficient in campaign? This is strange economy! To neglect to increase the soldier's pay five centimes! It would better his fare and prevent making of an officer a trader in vegetables in order to properly feed his men. Yet millions are squandered each year for uniforms, geegaws, shakos, etc!

If a big army is needed, it ought to cost as little as possible. Simplicity in all things! Down with all sorts of plumes! Less amateurs! If superfluous trimmings are not cut down it will be unfortunate! What is the matter with the sailor's uniform? Insignificant and annoying details abound while vital details of proper footgear and instruction, are neglected. The question of clothing for campaign is solved by adopting smocks and greatcoats and by doing away with headquarters companies! This is the height of folly. I suppose it is because our present uniforms need specialists to keep them in condition, and smocks and greatcoats do not!

APPENDIX I

MEMORANDUM ON INFANTRY FIRE

[Written in 1869 (Editor's note)]

1. Introduction

It may be said that the history of the development of infantry fire is none too plain, even though fire action to-day, in Europe, is almost the sole means of destruction used by that arm.

Napoleon said, "The only method of fire to be used in war is fire at will." Yet after such a plain statement by one who knew, there is a tendency to-day to make fire at command the basis of infantry battle tactics.

Is this correct? Experience only can determine. Experience is gained; but nothing, especially in the trade of war, is sooner forgotten than experience. So many fine things can be done, beautiful maneuvers executed, ingenious combat methods invented in the confines of an office or on the maneuver ground. Nevertheless let us try to hold to facts.

Let us consider, in the study of any kind of fire, a succinct history of small arms; let us see what kind of fire is used with each weapon, attempting at the same time to separate that which has actually happened from the written account.

2. Succinct History of the Development of Small Arms, from the Arquebus to Our Rifle

The arquebus in use before the invention of powder gave the general design to fire arms. The arquebus marks then the transition from the mechanically thrown missile to the bullet.

The tube was kept to direct the projectile, and the bow and string were replaced by a powder chamber and ignition apparatus.

This made a weapon, very simple, light and easy to charge; but the small caliber ball thrown from a very short barrel, gave penetration only at short distances.

The barrel was lengthened, the caliber increased, and a more efficient, but a less convenient arm resulted. It was indeed impossible to hold the weapon in aiming position and withstand the recoil at the moment of firing.

To lessen recoil there was attached to the bottom of the barrel a hook to catch on a fixed object at the moment of discharge. This was called a hook arquebus.

But the hook could only be used under certain circumstances. To give the arm a point of support on the body, the stock was lengthened and inclined to permit sighting. This was the petrinal or poitrinal. The soldier had in addition a forked support for the barrel.

In the musket, which followed, the stock was again modified and held against the shoulder. Further the firing mechanism was improved.

The arm had been fired by a lighted match; but with the musket, the arm becoming lighter and more portable, there came the serpentine lock, the matchlock, then the wheel-lock, finally the Spanish lock and the flint-lock.

The adoption of the flint-lock and the bayonet produced the rifle, which Napoleon regarded as the most powerful weapon that man possesses.

But the rifle in its primitive state had defects. Loading was slow; it was inaccurate, and under some circumstances it could not be fired.

How were these defects remedied?

As to the loading weakness, Gustavus Adolphus, understanding the influence on morale of rapid loading and the greater destruction caused by the more rapid fire, invented the cartridge for muskets. Frederick, or some one of his time, the name marks the period, replaced wooden by cylindrical iron ramrods. To prime more quickly a conical funnel allowed the powder to pass from the barrel into the firing-pan. These two last improvements saved time in two ways, in priming and in loading. But it was the adoption of the breech-loader that brought the greatest increase in rapidity of fire.

These successive improvements of the weapon, all tending to increase the rapidity of fire, mark the most remarkable military periods of modern times:

cartridges—Gustavus Adolphus
iron ramrod—Frederick
improved vent (adopted by the soldiers if not prescribed by competent
    orders)—wars of the Republic and of the Empire
breech-loading—Sadowa.

Accuracy was sacrificed to rapidity of fire. This will be explained later. Only in our day has the general use of rifling and of elongated projectiles brought accuracy to the highest point. In our times, also, the use of fulminate has assured fire under all conditions.

We have noted briefly the successive improvements in fire arms, from the arquebus to the rifle.

Have the methods of employment made the same progress?

3. Progressive Introduction of Fire-Arms Into the Armament of the Infantryman

The revolution brought about by powder, not in the art of war but in that of combat, came gradually. It developed along with the improvement of fire arms. Those arms gradually became those of the infantryman.

Thus, under Francis I, the proportion of infantrymen carrying fire arms to those armed with pikes was one to three or four.

At the time of the wars of religion arquebusiers and pikemen were about equal in number.

Under Louis XIII, in 1643, there were two fire-arms to one pike; in the war of 1688, four to one; finally pikes disappeared.

At first men with fire-arms were independent of other combatants, and functioned like light troops in earlier days.

Later the pikes and the muskets were united in constituent elements of army corps.

The most usual formation was pikes in the center, muskets on the wings.

Sometimes the pikemen were in the center of their respective companies, which were abreast.

Or, half the musketeers might be in front of the pikemen, half behind. Or again, all the musketeers might be behind the kneeling pikemen. In these last two cases fire covered the whole front.

Finally pike and musket might alternate.

These combinations are found in treatises on tactics. But we do not know, by actual examples, how they worked in battle, nor even whether all were actually employed.

4. The Classes of Fire Employed With Each Weapon

When originally some of the infantry were armed with the long and heavy arquebus in its primitive state, the feebleness of their fire caused Montaigne to say, certainly on military authority, "The arms have so little effect, except on the ears, that their use will be discontinued." Research is necessary to find any mention of their use in the battles of that period. [47]

However we find a valuable piece of information in Brantôme, writing of the battle of Pavia.

"The Marquis de Pescani won the battle of Pavia with Spanish arquebusiers, in an irregular defiance of all regulation and tradition by employing a new formation. Fifteen hundred arquebusiers, the ablest, the most experienced, the cleverest, above all the most agile and devoted, were selected by the Marquis de Pescani, instructed by him on new lines, and practiced for a long time. They scattered by squads over the battlefield, turning, leaping from one place to another with great speed, and thus escaped the cavalry charge. By this new method of fighting, unusual, astonishing, cruel and unworthy, these arquebusiers greatly hampered the operations of the French cavalry, who were completely lost. For they, joined together and in mass, were brought to earth by these few brave and able arquebusiers. This irregular and new method of fighting is more easily imagined than described. Any one who can try it out will find it is good and useful; but it is necessary that the arquebusiers be good troops, very much on the jump (as the saying is) and above all reliable."

It should be borne in mind, in noting the preceding, that there is always a great difference between what actually occurred, and the description thereof (made often by men who were not there, and God knows on what authority). Nevertheless, there appears in these lines of Brantôme a first example of the most destructive use of the rifle, in the hands of skirmishers.

During the religious wars, which consisted of skirmishes and taking and retaking garrisoned posts, the fire of arquebusiers was executed without order and individually, as above.

The soldier carried the powder charges in little metal boxes hung from a bandoleer. A finer, priming, powder was contained in a powder horn; the balls were carried in a pouch. At the onset the soldier had to load his piece. It was thus that he had to fight with the match arquebus. This was still far from fire at command.

However this presently appeared. Gustavus Adolphus was the first who tried to introduce method and coördination into infantry fire. Others, eager for innovations, followed in his path. There appeared successively, fire by rank, in two ranks, by subdivision, section, platoon, company, battalion, file fire, parapet fire, a formal fire at will, and so many others that we can be sure that all combinations were tried at this time.

Fire by ranks was undoubtedly the first of these; it will give us a line on the others.

Infantry was formed six deep. To execute fire by rank all ranks except the last knelt. The last rank fired and reloaded. The rank in front of it then rose and did the same thing, as did all other ranks successively. The whole operation was then recommenced.

Thus the first group firing was executed successively by ranks.

Montecuculli said, "The musketeers are ranged six deep, so that the last rank has reloaded by the time the first has fired, and takes up the fire again, so that the enemy has to face continuous fire."

However, under Condé and Turenne, we see the French army use only fire at will.

It is true that at this time fire was regarded only as an accessory. The infantry of the line which, since the exploit of the Flemish, the Swiss and the Spaniards, had seen their influence grow daily, was required for the charge and the advance and consequently was armed with pikes.

In the most celebrated battles of these times, Rocroi, Nordlingen, Lens, Rethel and the Dunes, we see the infantry work in this way. The two armies, in straight lines, commenced by bombarding each other, charged with their cavalry wings, and advanced with their infantry in the center. The bravest or best disciplined infantry drove back the other, and often, if one of its wings was victorious, finished by routing it. No marked influence of fire is found at this time. The tradition of Pescani was lost.

Nevertheless fire-arms improved; they became more effective and tended to replace the pike. The use of the pike obliged the soldier to remain in ranks, to fight only in certain cases, and exposed him to injury without being able to return blow for blow. And, this is exceedingly instructive, the soldier had by this time an instinctive dislike of this arm, which often condemned him to a passive role. This dislike necessitated giving high pay and privilege to obtain pikemen. And in spite of all at the first chance the soldier threw away his pike for a musket.

The pikes themselves gradually disappeared before firearms; the ranks thinned to permit the use of the latter. Four rank formation was used, and fire tried in that order, by rank, by two ranks, upright, kneeling, etc.

In spite of these attempts, we see the French army in combat, notably at Fontenoy, still using fire at will, the soldier leaving ranks to fire and returning to load.

It can be stated, in spite of numerous attempts at adoption, that no fire at command was used in battle up to the days of Frederick.

Already, under William, the Prussian infantry was noted for the rapidity and continuity of its fire. Frederick further increased the ability of his battalions to fire by decreasing their depth. This fire, tripled by speed in loading, became so heavy that it gave Prussian battalions a superiority over others of three to one.

The Prussians recognized three kinds of fire, at a halt, in advancing, and in retreat. We know the mechanics of fire at a halt, the first rank kneeling. Of fire in advancing Guibert says: "What I call marching fire, and which anybody who thinks about it must find as ill advised as I do, is a fire I have seen used by some troops. The soldiers, in two ranks, fire in marching, but they march of course at a snail's pace. This is what Prussian troops call fire in advancing. It consists in combined and alternating volleys from platoons, companies, half battalions or battalions. The parts of the line which have fired advance at the double, the others at the half step."

In other methods of fire, as we have said, the Prussian battalion was in three ranks; the first kneeling. The line delivered salvos, only at command.

However, the theory of executing fire by salvo in three ranks did not bother Frederick's old soldiers. We will see presently how they executed it on the field of battle.

Be that as it may, Europe was impressed with these methods and tended to adopt them. D'Argenson provided for them in the French army and introduced fire at command. Two regulations prescribing this appeared, in 1753 and 1755. But in the war which followed, Marshal de Broglie, who undoubtedly had experience and as much common sense as M. D'Argenson, prescribed fire at will. All infantry in his army was practiced in it during the winter of 1761-1762.

Two new regulations succeeded the preceding, in 1764 and 1776. The last prescribed fire in three ranks at command, all ranks upright. [48]

Thus we come to the wars of the Revolution, with regulations calling for fire at command, which was not executed in battle.

Since these wars, our armies have always fought as skirmishers. In speaking of our campaigns, fire at command is never mentioned. It was the same under the Empire, in spite of numerous essays from the Boulogne school and elsewhere. At the Boulogne school, fire at command by ranks was first tried by order of Napoleon. This fire, to be particularly employed against cavalry—in theory it is superb—does not seem to have been employed Napoleon says so himself, and the regulations of 1832, in which some influence of soldiers of the Empire should be found, orders fire in two ranks or at will, by bodies of men, to the exclusion of all others.

According to our military authority, on the authority of our old officers, fire at command did not suit our infantry; yet it lived in the regulations. General Fririon (1822) and de Gouvion-Saint-Cyr (1829) attacked this method. Nothing was done. It remained in the regulations of 1832, but without being ordered in any particular circumstances. It appeared there for show purposes, perhaps.

On the creation of the chasseurs d'Orléans, fire by rank was revived. But neither in our African campaigns nor in our last two wars in the Crimea and Italy can a single example of fire at command be found. In practice it was believed to be impracticable. It was known to be entirely ineffective and fell into disrepute.

But to-day, with the breech-loading rifle, there is a tendency to believe it practicable and to take it up with new interest. Is this more reasonable than in the past? Let us see.

5. Methods of Fire Used in the Presence of the Enemy; Methods Recommended or Ordered But Impractical. Use and Efficacy of Fire at Command

Undoubtedly at the Potsdam maneuvers the Prussian infantry used only salvos executed admirably. An unbelievable discipline kept the soldier in place and in line. Barbaric punishments were incorporated in the military code. Blows, the whip, executions, punished the slightest derelictions. Even N.C.O.'s were subjected to blows with the flat of the sword. Yet all this was not enough on the field of battle; a complete rank of non-commissioned officer file closers was also needed to hold the men to their duty.

M. Carion-Nisas said, "These file-closers hook their halberds together and form a line that cannot be broken." In spite of all this, after two or three volleys, so says General Renard, whom we believe more than charitable, there is no power of discipline which can prevent regular fire from breaking into fire at will.

But let us look further, into Frederick's battles. Let us take the battle of Mollwitz, in which success was specifically laid to fire at command, half lost, then won by the Prussian salvos.

"The Austrian infantry had opened fire on the lines of the Prussians, whose cavalry had been routed. It was necessary to shake them to insure victory. The Austrians still used wooden ramrods. Their fire came slowly, while the Prussian fire was thunderous, five or six shots to the rifle per minute. The Imperial troops, surprised and disconcerted by this massed fire, tried to hurry. In their hurry many broke their fragile ramrods. Confusion spread through the ranks, and the battle was lost."

But, if we study actual conditions of the period, we see that things did not happen in such an orderly sequence.

Firing started, and it is said that it was long and deadly. The Prussians iron ramrods gave them the advantage 'over an enemy whose ramrods were wooden, harder to manipulate and easily broken. However, when the order to advance was given to the Prussians, whole battalions stood fast; it was impossible to budge them. The soldiers tried to escape the fire and got behind each other, so that they were thirty to forty deep.

Here are men who exhibit under fire an admirable, calm, an immovable steadiness. Each instant they hear the dead heavy sound of a bullet striking. They see, they feel, around them, above them, between their legs, their comrades fall and writhe, for the fire is deadly. They have the power in their hands to return blow for blow, to send back to the enemy the death that hisses and strikes about them. They do not take a false step; their hands do not close instinctively on the trigger. They wait, imperturbably, the order of their chiefs—and what chiefs! These are the men who at the command "forward," lack bowels, who huddle like sheep one behind the other. Are we to believe this?

Let us get to the truth of the matter. Frederick's veterans, in spite of their discipline and drill, are unable to follow the methods taught and ordered. They are no more able to execute fire at command than they are to execute the ordered advance of the Potsdam maneuver field. They use fire at will. They fire fast from instinct—stronger than their discipline—which bids them send two shots for one. Their fire becomes indeed, a thunderous roll, not of salvos, but of rapid fire at will. Who fires most, hits most, so the soldier figures. So indeed did Frederick, for he encouraged fire in this same battle of Mollwitz; he thereafter doubled the number of cartridges given the soldier, giving him sixty instead of thirty.

Furthermore, if fire at command had been possible, who knows what Frederick's soldiers would have been capable of? They would have cut down battalions like standing grain. Allowed to aim quietly, no man interfering with another, each seeing clearly—then at the signal all firing together. Could anything hold against them? At the first volley the enemy would have broken and fled, under the penalty of annihilation in case they stayed. However, if we look at the final result at Mollwitz, we see that the number of killed is about the same on the side that used fire at command as on the side that did not. The Prussians lost 960 dead, the Austrians 966.

But they say that if fire was not more deadly, it was because sight-setting was then unknown. What if it was? There was no adjustment of fire perhaps, but there were firing regulations; aiming was known. Aiming is old. We do not say it was practiced; but it was known, and often mentioned. Cromwell often said, "Put your confidence in God, my children, and fire at their shoe-laces."

Do we set our sights better to-day? It is doubtful. If the able soldiers of Cromwell, of Frederick, of the Republic and of Napoleon could not set their sights—can we?

Thus this fire at command, which was only possible rarely and to commence action, was entirely ineffective.

Hardy spirits, seeing the slight effect of long range firing in battle, counselled waiting till the enemy was at twenty paces and driving him back with a volley. You do not have to sight carefully at twenty paces. What would be the result?

"At the battle of Castiglione," says Marshal Saxe, "the Imperial troops let the French approach to twenty paces, hoping to destroy them by a volley. At that distance they fired coolly and with all precautions, but they were broken before the smoke cleared. At the battle of Belgrade (1717) I saw two battalions who at thirty paces, aimed and fired at a mass of Turks. The Turks cut them up, only two or three escaping. The Turkish loss in dead was only thirty-two."

No matter what the Marshal says, we doubt that these men were cool. For men who could hold their fire up to such a near approach of the enemy, and fire into masses, would have killed the front rank, thrown the others into confusion, and would never have been cut up as they were. To make these men await, without firing, an enemy at twenty or thirty paces, needed great moral pressure. Controlled by discipline they waited, but as one waits for the roof to fall, for a bomb to explode, full of anxiety and suppressed emotion. When the order is given to raise the arms and fire the crisis is reached. The roof falls, the bomb explodes, one flinches and the bullets are fired into the air. If anybody is killed it is an accident.

This is what happened before the use of skirmishers. Salvos were tried. In action they became fire at will. Directed against troops advancing without firing they were ineffective. They did not halt the dash of the assault, and the troops who had so counted on them fled demoralized. But when skirmishers were used, salvos became impossible. Armies who held to old methods learned this to their cost.

In the first days of the Revolution our troops, undrilled and not strictly disciplined, could not fight in line. To advance on the enemy, a part of the battalion was detached as skirmishers. The remainder marched into battle and was engaged without keeping ranks. The combat was sustained by groups fighting without formal order. The art was to support by reserves the troops advanced as skirmishers. The skirmishers always began the action, when indeed they did not complete it.

To oppose fire by rank to skirmishers was fools' play.

Skirmishers necessarily opposed each other. Once this method was adopted, they were supported, reinforced by troops in formation. In the midst of general firing fire at command became impossible and was replaced by fire at will.

Dumouriez, at the battle of Jemmapes, threw out whole battalions as skirmishers, and supporting them by light cavalry, did wonders with them. They surrounded the Austrian redoubts and rained on the cannoneers a hail of bullets so violent that they abandoned their pieces.

The Austrians, astounded by this novel combat method, vainly reinforced their light troops by detachments of heavy infantry. Their skirmishers could not resist our numbers and impetuosity, and presently their line, beaten by a storm of bullets, was forced back. The noise of battle, the firing, increased; the defeated troops, hearing commands no longer, threw down their arms and fled in disorder.

So fire in line, heavy as it may be, cannot prevail against the power of numerous detachments of skirmishers. A rain of bullets directed aimlessly is impotent against isolated men profiting by the slightest cover to escape the fire of their adversaries, while the deployed battalions offer to their rifles a huge and relatively harmless target. The dense line, apparently so strong, withers under the deadly effect of the fire of isolated groups, so feeble in appearance. (General Renard.)

The Prussians suffered in the same way at Jena. Their lines tried fire at command against our skirmishers. You might as well fire on a handful of fleas.

They tell us of the English salvos at Sainte-Euphémie, in Calabria, and later in Spain. In these particular cases they could be used, because our troops charged without first sending out skirmishers.

The battle of Sainte-Euphémie only lasted half an hour; it was badly conceived and executed, "And if," says General Duhesme, "the advancing battalions had been preceded by detachments of skirmishers who had already made holes in enemy ranks, and, on close approach, the heads of columns had been launched in a charge, the English line would not have conserved that coolness which made their fire so effective and accurate. Certainly it would not have waited so long to loose its fire, if it had been vigorously harassed by skirmishers."

An English author, treating of the history of weapons, speaks of the rolling fire, well directed, of the English troops. He makes no mention of salvos. Perhaps we were mistaken, and in our accounts have taken the fire of a battalion for the formal battalion fire at command of our regulations.

The same tendency appears more clearly in the work on infantry of the Marquis de Chambray, who knew the English army well. He says that the English in Spain used almost entirely fire in two ranks. They employed battalion fire only when attacked by our troops without skirmishers, firing on the flanks of our columns. And he says "The fire by battalion, by half battalion and by platoon is limited to the target range. The fire actually most used in war is that in two ranks, the only one used by the French." Later he adds "Experience proves fire in two ranks the only one to be used against the enemy." Before him Marshal Saxe wrote "Avoid dangerous maneuvers, such as fire by platoon, which have often caused shameful defeats." These statements are as true now as then.

Fire at command, by platoon, by battalion, etc., is used in case the enemy having repulsed skirmishers and arrived at a reasonable range either charges or opens fire for effect himself. If the latter, fire is reciprocal and lasts until one or the other gives way or charges. If the enemy charges, what happens? He advances preceded by skirmishers who deliver a hail of bullets. You wish to open fire, but the voices of your officers are lost. The noise of artillery, of small arms, the confusion of battle, the shrieks of the wounded, distract the soldiers' attention. Before you have delivered your command the line is ablaze. Then try to stop your soldiers. While there is a cartridge left, they will fire. The enemy may find a fold of ground that protects him; he may adopt in place of his deployed order columns with wide intervals between, or otherwise change his dispositions. The changing incidents of battle are hidden by smoke and the troops in front, from the view of the officers behind. The soldiers will continue to fire and the officers can do nothing about it.

All this has been said already, has been gone into, and fire at command has been abandoned. Why take it up again? It comes to us probably from the Prussians. Indeed the reports of their general staff on their last campaign, of 1866, say that it was very effectively employed, and cite many examples.

But a Prussian officer who went through the campaign in the ranks and saw things close up, says, "In examining the battles of 1866 for characteristics, one is struck by a feature common to all, the extraordinary extension of front at the expense of depth. Either the front is spun out into a single long thin line, or it is broken into various parts that fight by themselves. Above all the tendency is evident to envelop the enemy by extending the wings. There is no longer any question of keeping the original order of battle. Different units are confused, by battle, or even before battle. Detachments and large units of any corps are composed of diverse and heterogeneous elements. The battle is fought almost exclusively by columns of companies, rarely of half-battalions. The tactics of these columns consists in throwing out strong detachments of skirmishers. Gradually the supports are engaged and deployed. The line is broken, scattered, like a horde of irregular cavalry. The second line which has held close order tries to get up to the first promptly, first to engage in the fight, also because they suffer losses from the high shots directed at the first line. It suffers losses that are heavy as it is compact and supports them with impatience as it does not yet feel the fever of battle. The most of the second line then forces entry into the first, and, as there is more room on the wings, it gravitates to the wings. Very often even the reserve is drawn in, entirely, or so largely that it cannot fulfill its mission. In fact, the fighting of the first two lines is a series of combats between company commands and the enemy each command faces. Superior officers cannot follow on horseback all the units, which push ahead over all sorts of ground. They have to dismount and attach themselves to the first unit of their command met. Unable to manipulate their whole command, in order to do something, they command the smaller unit. It is not always better commanded at that. Even generals find themselves in this situation."

Here is something we understand better. It is certainly what occurs.

As for the instances cited in the general staff reports, they deal with companies or half-battalions at most. Not withstanding the complacency with which they are cited, they must have been rare, and the exception should not be taken as establishing a rule.

6. Fire at Will—Its Efficacy

Thus fire at command, to-day as in the past, is impractical and consequently not actually used in battle. The only means employed are fire at will and the fire of skirmishers. Let us look into their efficacy.

Competent authorities have compiled statistics on this point.

Guibert thinks that not over two thousand men are killed or wounded by each million cartridges used in battle.

Gassendi assures us that of three thousand shots only one is a hit.

Piobert says that the estimate, based on the result of long wars, is that three to ten thousand cartridges are expended for each man hit.

To-day, with accurate and long range weapons, have things changed much? We do not think so. The number of bullets fired must be compared with the number of men dropped, with a deduction made for the action of artillery, which must be considered.

A German author has advanced the opinion that with the Prussian needle rifle the hits are 60% of the shots fired. But then how explain the disappointment of M. Dreyse, the happy inventor of the needle rifle, when he compared Prussian and Austrian losses. This good old gentleman was disagreeably astonished at seeing that his rifle had not come up to his expectations.

Fire at will, as we shall presently show, is a fire to occupy the men in the ranks but its effect is not great. We could give many examples; we only cite one, but it is conclusive.

"Has it not been remarked," says General Duhesme, "that, before a firing line there is raised a veil of smoke which on one side or the other hides the troops from view, and makes the fire of the best placed troops uncertain and practically without effect? I proved it conclusively at the battle of Caldiero, in one of the successive advances that occurred on my left wing. I saw some battalions, which I had rallied, halted and using an individual fire which they could not keep up for long. I went there. I saw through the smoke cloud nothing but flashes, the glint of bayonets and the tops of grenadier's caps. We were not far from the enemy however, perhaps sixty paces. A ravine separated us, but it could not be seen. I went into the ranks, which were neither closed nor aligned, throwing up with my hand the soldiers' rifles to get them to cease firing and to advance. I was mounted, followed by a dozen orderlies. None of us were wounded, nor did I see an infantryman fall. Well then! Hardly had our line started when the Austrians, heedless of the obstacle that separated us, retreated."

It is probable that had the Austrians started to move first, the French would have given way. It was veterans of the Empire, who certainly were as reliable as our men, who gave this example of lack of coolness.

In ranks, fire at will is the only possible one for our officers and men. But with the excitement, the smoke, the annoying incidents, one is lucky to get even horizontal fire, to say nothing of aimed fire.

In fire at will, without taking count of any trembling, men interfere with each other. Whoever advances or who gives way to the recoil of his weapon deranges the shot of his neighbor. With full pack, the second rank has no loophole; it fires in the air. On the range, spacing men to the extremity of the limits of formation, firing very slowly, men are found who are cool and not too much bothered by the crack of discharge in their ears, who let the smoke pass and seize a loophole of pretty good visibility, who try, in a word, not to lose their shots. And the percentage results show much more regularity than with fire at command.

But in front of the enemy fire at will becomes in an instant haphazard fire. Each man fires as much as possible, that is to say, as badly as possible. There are physical and mental reasons why this is so.

Even at close range, in battle, the cannon can fire well. The gunner, protected in part by his piece, has an instant of coolness in which to lay accurately. That his pulse is racing does not derange his line of sight, if he has will power. The eye trembles little, and the piece once laid, remains so until fired.

The rifleman, like the gunner, only by will-power keeps his ability to aim. But the excitement in the blood, of the nervous system, opposes the immobility of the weapon in his hands. No matter how supported, a part of the weapon always shares the agitation of the man. He is instinctively in haste to fire his shot, which may stop the departure of the bullet destined for him. However lively the fire is, this vague reasoning, unformed as it is in his mind, controls with all the force of the instinct of self preservation. Even the bravest and most reliable soldiers then fire madly.

The greater number fire from the hip.

The theory of the range is that with continual pressure on the trigger the shot surprises the firer. But who practices it under fire?

However, the tendency in France to-day is to seek only accuracy. What good will it do when smoke, fog, darkness, long range, excitement, the lack of coolness, forbid clear sight?

It is hard to say, after the feats of fire at Sebastopol, in Italy, that accurate weapons have given us no more valuable service than a simple rifle. Just the same, to one who has seen, facts are facts. But—see how history is written. It has been set down that the Russians were beaten at Inkermann by the range and accuracy of weapons of the French troops. But the battle was fought in thickets and wooded country, in a dense fog. And when the weather cleared, our soldiers, our chasseurs were out of ammunition and borrowed from the Russian cartridge boxes, amply provided with cartridges for round, small calibered bullets. In either case there could have been no accurate fire. The facts are that the Russians were beaten by superior morale; that unaimed fire, at random, there perhaps more than elsewhere, had the only material effect.

When one fires and can only fire at random, who fires most hits most. Or perhaps it is better said that who fires least expects to be hit most.

Frederick was impressed with this, for he did not believe in the Potsdam maneuvers. The wily Fritz looked on fire as a means to quiet and occupy the undependable soldiers and it proved his ability that he could put into practice that which might have been a mistake on the part of any other general officer. He knew very well how to count on the effect of his fire, how many thousand cartridges it took to kill or wound an enemy. At first his soldiers had only thirty cartridges. He found the number insufficient, and after Mollwitz gave them sixty.

To-day as in Frederick's day, it is rapid random fire, the only one practicable, which has given prestige to the Prussians. This idea of rapid fire was lost after Frederick, but the Prussians have recovered it to-day by exercising common sense. However our veterans of the Empire had preserved this idea, which comes from instinct. They enlarged their vents, scornful of flare backs, to avoid having to open the chamber and prime. The bullet having a good deal of clearance when the cartridge was torn and put in the gun, with a blow of the butt on the ground they had their arms charged and primed.

But to-day as then, in spite of skill acquired in individual fire, men stop aiming and fire badly as soon as they are grouped into platoons to fire.

Prussian officers, who are practical men, know that adjustment of sights is impracticable in the heat of action, and that in fire by volleys troops tend to use the full sight. So in the war of 1866 they ordered their men to fire very low, almost without sighting, in order to profit by ricochets.

7. Fire by Rank Is a Fire to Occupy the Men in Ranks

But if fire at will is not effective, what is its use? As we have already said its use is to occupy the men in the ranks.

In ordinary fire the act of breathing alone, by the movement it communicates to the body greatly annoys men in firing. How then can it be claimed that on the field of battle, in rank, men can fire even moderately well when they fire only to soothe themselves and forget danger?

Napoleon said "The instinct of man is not to let himself be killed without defending himself." And indeed man in combat is a being in whom the instinct of self preservation dominates at times all other sentiments. The object of discipline is to dominate this instinct by a greater terror of shame or of punishment. But it is never able entirely to attain this object; there is a point beyond which it is not effectual. This point reached, the soldier must fire or he will go either forward or back. Fire is then, let us say, a safety vent for excitement.

In serious affairs it is then difficult, if not impossible, to control fire. Here is an example given by Marshal Saxe:

"Charles XII, King of Sweden, wished to introduce into his infantry the method of charging with the bayonet. He spoke of it often, and it was known in the army that this was his idea. Finally at the battle of —— against the Russians, when the fighting started he went to his regiment of infantry, made it a fine speech, dismounted before the colors, and himself led the regiment to the charge. When he was thirty paces from the enemy the whole regiment fired, in spite of his orders and his presence. Otherwise, it did very well and broke the enemy. The king was so annoyed that all he did was pass through the ranks, remount his horse, and go away without saying a word."

So that, if the soldier is not made to fire, he will fire anyway to distract himself and forget danger. The fire of Frederick's Prussians had no other purpose. Marshal Saxe saw this. "The speed with which the Prussians load their rifles," he tells us, "is advantageous in that it occupies the soldier and forbids reflection while he is in the presence of the enemy. It is an error to believe that the five last victories gained by the nation in its last war were due to fire. It has been noted that in most of these actions there were more Prussians killed by rifle fire than there were of their enemies."

It would be sad to think the soldier in line a firing machine. Firing has been and always will be his principal object, to fire as many shots in as short a time as possible. But the victor is not always the one who kills the most; he is fortunate who best knows how to overcome the morale of his enemy.

The coolness of men cannot be counted on. And as it is necessary above all to keep up their morale one ought to try above all to occupy and soothe them. This can best be done by frequent discharges. There will be little effect, and it would be absurd to expect them to be calm enough to fire slowly, adjust their ranges and above all sight carefully.

8. The Deadly Fire Is the Fire of Skirmishers

In group firing, when the men are grouped into platoons or battalions, all weapons have the same value, and if it is assumed to-day that fire must decide engagements, the method of fighting must be adopted which gives most effect to the weapon. This is the employment of skirmishers.

It is this class of fire, indeed, which is deadliest in war. We could give many examples but we shall be content with the two following instances, taken from General Duhesme.

"A French officer who served with the Austrians in one of the recent wars," says General Duhesme, "told me that from the fire of a French battalion one hundred paces from them, his company lost only three or four men, while in the same time they had had more than thirty killed or wounded by the fire of a group of skirmishers in a little wood on their flank three hundred paces away."

"At the passage of the Minico, in 1801, the 2nd battalion of the 91st received the fire of a battalion of Bussi's regiment without losing a man; the skirmishers of that same organization killed more than thirty men in a few minutes while protecting the retreat of their organization."

The fire of skirmishers is then the most deadly used in war, because the few men who remain cool enough to aim are not otherwise annoyed while employed as skirmishers. They will perform better as they are better hidden, and better trained in firing.

The accuracy of fire giving advantages only in isolated fire, we may consider that accurate weapons will tend to make fighting by skirmishers more frequent and more decisive.

For the rest, experience authorizes the statement that the use of skirmishers is compulsory in war. To-day all troops seriously engaged become in an instant groups of skirmishers and the only possible precise fire is from hidden snipers.

However, the military education which we have received, the spirit of the times, clouds with doubt our mind regarding this method of fighting by skirmishers. We accept it regretfully. Our personal experience being incomplete, insufficient, we content ourselves with the supposition that gives us satisfaction. The war of skirmishers, no matter how thoroughly it has been proven out, is accepted by constraint, because we are forced by circumstance to engage our troops by degrees, in spite of ourselves, often unconsciously. But, be it understood, to-day a successive engagement is necessary in war.

However, let us not have illusions as to the efficacy of the fire of skirmishers. In spite of the use of accurate and long range weapons, in spite of all training that can be given the soldier, this fire never has more than a relative effect, which should not be exaggerated.

The fire of skirmishers is generally against skirmishers. A body of troops indeed does not let itself be fired on by skirmishers without returning a similar fire. And it is absurd to expect skirmishers to direct their fire on a body protected by skirmishers. To demand of troops firing individually, almost abandoned to themselves, that they do not answer the shots directed at them, by near skirmishers, but aim at a distant body, which is not harming them, is to ask an impossible unselfishness.

As skirmishers men are very scattered. To watch the adjustment of ranges is difficult. Men are practically left alone. Those who remain cool may try to adjust their range, but it is first necessary to see where your shots fall, then, if the terrain permits this and it will rarely do so, to distinguish them from shots fired at the same time by your neighbors. Also these men will be more disturbed, will fire faster and less accurately, as the fight is more bitter, the enemy stauncher; and perturbation is more contagious than coolness.

The target is a line of skirmishers, a target offering so little breadth and above all depth, that outside of point blank fire, an exact knowledge of the range is necessary to secure effect. This is impossible, for the range varies at each instant with the movements of the skirmishers. [49]

Thus, with skirmishers against skirmishers, there are scattered shots at scattered targets. Our fire of skirmishers, marching, on the target range, proves this, although each man knows exactly the range and has time and the coolness to set his sights. It is impossible for skirmishers in movement to set sights beyond four hundred meters, and this is pretty extreme, even though the weapon is actually accurate beyond this.

Also, a shot is born. There are men, above all in officer instructors at firing schools, who from poor shots become excellent shots after years of practice. But it is impossible to give all the soldiers such an education without an enormous consumption of ammunition and without abandoning all other work. And then there would be no results with half of them.

To sum up, we find that fire is effective only at point blank. Even in our last wars there have been very few circumstances in which men who were favored with coolness and under able leadership have furnished exceptions. With these exceptions noted, we can say that accurate and long range weapons have not given any real effect at a range greater than point blank.

There has been put forward, as proof of the efficacy of accurate weapons the terrible and decisive results obtained by the British in India, with the Enfield rifle. But these results have been obtained because the British faced comparatively poorly armed enemies. They had then the security, the confidence, the ensuing coolness necessary for the use of accurate weapons. These conditions are completely changed when one faces an enemy equally well armed, who consequently, gives as good as he gets.

9. Absolute Impossibility of Fire at Command

Let us return to fire at command, which there is a tendency to-day to have troops execute in line.

Can regular and efficient fire be hoped for from troops in line? Ought it to be hoped for?

No, for man cannot be made over, and neither can the line.

Even on the range or on the maneuver field what does this fire amount to?

In fire at command, on the range, all the men in the two ranks come to the firing position simultaneously, everybody is perfectly quiet. Men in the front rank consequently are not deranged by their neighbors. Men in the second rank are in the same situation. The first rank being set and motionless they can aim through the openings without more annoyance than those in the first rank.

Fire being executed at command, simultaneously, no weapon is deranged at the moment of firing by the movements of the men. All conditions are entirely favorable to this kind of fire. Also as the fire is ordered with skill and coolness by an officer who has perfectly aligned his men (a thing rare even on the drill ground) it gives percentage results greater than that of fire at will executed with the minutest precautions, results that are sometimes astonishing.

But fire at command, from the extreme coolness that it demands of all, of the officer certainly more than of the soldier, is impracticable before the enemy except under exceptional circumstances of picked officers, picked men, ground, distance, safety, etc. Even in maneuvers its execution is farcical. There is not an organization in which the soldiers do not hurry the command to fire in that the officers are so afraid that their men will anticipate the command that they give it as rapidly as possible, while the pieces are hardly in firing position, often while they are still in motion.

The prescription that the command to fire be not given until about three seconds after coming to the firing position may give good results in the face of range targets. But it is not wise to believe that men will wait thus for long in the face of the enemy.

It is useless to speak of the use of the sight-leaf before the enemy, in fire attempted by the same officers and men who are so utterly lacking, even on the maneuver ground. We have seen a firing instructor, an officer of coolness and assurance, who on the range had fired trial shots every day for a month, after this month of daily practice fire four trial shots at a six hundred meter range with the sight leaf at point blank.

Let us not pay too much attention to those who in military matters base everything on the weapon and unhesitating assume that the man serving it will adopt the usage provided and ordered in their regulations. The fighting man is flesh and blood. He is both body and soul; and strong as the soul may often be it cannot so dominate the body that there is no revolt of the flesh, no mental disturbance, in the face of destruction. Let us learn to distrust mathematics and material dynamics as applied to battle principles. We shall learn to beware of the illusions drawn from the range and the maneuver field.

There experience is with the calm, settled, unfatigued, attentive, obedient soldier, with an intelligent and tractable man instrument in short. And not with the nervous, easily swayed, moved, troubled, distrait, excited, restless being, not even under self-control, who is the fighting man from general to private. There are strong men, exceptions, but they are rare.

These illusions nevertheless, stubborn and persistent, always repair the next day the most damaging injuries inflicted on them by reality. Their least dangerous effect is to lead to prescribing the impracticable, as if ordering the impracticable were not really an attack on discipline, and did not result in disconcerting officers and men by the unexpected and by surprise at the contrast between battle and the theories of peace-time training.

Battle of course always furnishes surprises. But it furnishes less in proportion as good sense and the recognition of the truth have had their effect on the training of the fighting man.

Man in the mass, in a disciplined body organized for combat, is invincible before an undisciplined body. But against a similarly disciplined body he reverts to the primitive man who flees before a force that is proved stronger, or that he feels stronger. The heart of the soldier is always the human heart. Discipline holds enemies face to face a little longer, but the instinct of self-preservation maintains its empire and with it the sense of fear.

Fear!

There are chiefs, there are soldiers who know no fear, but they are of rare temper. The mass trembles, for the flesh cannot be suppressed. And this trembling must be taken into account in all organization, discipline, formation, maneuver, movement, methods of action. For in all of these the soldier tends to be upset, to be deceived, to under-rate himself and to exaggerate the offensive spirit of the enemy.

On the field of battle death is in the air, blind and invisible, making his presence known by fearful whistlings that make heads duck. During this strain the recruit hunches up, closes in, seeking aid by an instinctive unformulated reasoning. He figures that the more there are to face a danger the greater each one's chances of escaping. But he soon sees that flesh attracts lead. Then, possessed by terror, inevitably he retreats before the fire, or "he escapes by advancing," in the picturesque and profound words of General Burbaki.

The soldier escapes from his officer, we say. Yes, he escapes! But is it not evident that he escapes because up to this moment nobody has bothered about his character, his temperament, the impressionable and exciteable nature of man? In prescribed methods of fighting he has always been held to impossibilities. The same thing is done to-day. To-morrow, as yesterday, he will escape.

There is of course a time when all the soldiers escape, either forward, or to the rear. But the organization, the combat methods should have no other object than to delay as long as possible this crisis. Yet they hasten it.

All our officers fear, quite justifiably from their experience, that the soldier will too rapidly use his cartridges in the face of the enemy. This serious matter is certainly worthy of attention. How to stop this useless and dangerous waste of ammunition is the question. Our soldiers show little coolness. Once in danger they fire, fire to calm themselves, to pass the time; they cannot be stopped.

There are some people you cannot embarrass. With the best faith in the world they say, "What is this? You are troubled about stopping the fire of your soldiers? That is not difficult. You find that they show little coolness, and shoot despite their officers, in spite even of themselves? All right, require of them and their officers methods of fire that demand extremes of coolness, calm and assurance, even in maneuver. They cannot give a little? Ask a lot and you will get it. There you have a combat method nobody has ever heard of, simple, beautiful, and terrible."

This is indeed a fine theory. It would make the wily Frederick who surely did not believe in these maneuvers, laugh until he cried. [50]

This is to escape from a difficulty by a means always recognized as impossible, and more impossible than ever to-day.

Fearing that the soldier will escape from command, can not better means be found to hold him than to require of him and his officer, impracticable fire? This, ordered and not executed by the soldiers, and even by the officers, is an attack on the discipline of the unit. "Never order the impossible," says discipline, "for the impossible becomes then a disobedience."

How many requisites there are to make fire at command possible, conditions among the soldiers, among their officers. Perfect these conditions, they say. All right, perfect their training, their discipline, etc.; but to obtain fire at command it is necessary to perfect their nerves, their physical force, their moral force, to make bronze images of them, to do away with excitement, with the trembling of the flesh. Can any one do this?

Frederick's soldiers were brought, by blows of the baton, to a terrible state of discipline. Yet their fire was fire at will. Discipline had reached its limits.

Man in battle, let us repeat again, is a being to whom the instinct of self-preservation at times dominates everything else. Discipline, whose purpose is to dominate this instinct by a feeling of greater terror, can not wholly achieve it. Discipline goes so far and no farther.

We cannot deny the existence of extraordinary instances when discipline and devotion have raised man above himself. But these examples are extraordinary, rare. They are admired as exceptions, and the exception proves the rule.

As to perfection, consider the Spartans. If man was ever perfected for war it was he; and yet he has been beaten, and fled.

In spite of training, moral and physical force has limits. The Spartans, who should have stayed to the last man on the battle field, fled.

The British with their phlegmatic coolness and their terrible rolling fire, the Russians, with that inertia that is called their tenacity, have given way before attack. The German has given way, he who on account of his subordination and stability has been called excellent war material.

Again an objection is raised. Perhaps with recruits the method may be impracticable. But with veterans—But with whom is war commenced? Methods are devised precisely for young and inexperienced troops.

They ask, also, if the Prussians used this method of fire successfully in the last war, why should not we do as well? Supposing that the Prussians actually did use it, and this is far from being proved, it does not follow that it is practicable for us. This mania for borrowing German tactics is not new, although it has always been properly protested against. Marshal Luchner said, "No matter how much they torment their men, fortunately they will never make them Prussians." Later de Gouvion-Saint-Cyr said, "The men are drilled in various exercises believed necessary to fit them for war, but there is no question of adopting exercises to suit the French military genius, the French character and temperament. It has not been thought necessary to take this into account; it has been easier to borrow German methods."

To follow preconceived tactics is more the part of the phlegmatic German than it is ours. The Germans obey well enough, but the point is that they try to follow tactics which are contrary to nature. The Frenchman cannot. More spontaneous, more exciteable and impressionable, less calm and obedient, he has in our last wars promptly and completely violated both the letter and the spirit of the regulations. "The German," said a Prussian officer, "has sentiments of duty and obedience. He submits to severe discipline. He is full of devotion, although not animated by a lively mind. Easy by nature, rather heavy than active, intellectually calm, reflective, without dash or divine fire, wishing but not mad to conquer, obeying calmly and conscientiously, but mechanically and without enthusiasm, fighting with a resigned valor, with heroism, he may let himself be sacrificed uselessly, but he sells his life dearly. Without warlike tendencies, not bellicose, unambitious, he is yet excellent war material on account of his subordination and stability. What must be inculcated in him is a will of his own, a personal impulse to send him forward." According to this unflattering portrait, which we believe a little extreme, even if by a compatriot, it is possible that the Germans can be handled in tactics impossible with French. However, did they actually use these tactics? Remember the urgent warning of Blücher to his brigade commanders, not to let bayonet attacks break down into fusillades. Note the article in the present Prussian firing regulations, which prescribes trial shots before each fire delivered, "so as to dissipate the kind of excitement that possesses the soldier when his drill has been interrupted for some time."

In conclusion, if fire at command was impossible with the ancient rifle, it is more so to-day, for the simple reason that trembling increases as the destructive power increases. Under Turenne, lines held longer than to-day, because the musket was in use and the battle developed more slowly. To-day when every one has the rapid fire rifle, are things easier? Alas no! Relations between weapons and the man are the same. You give me a musket, I fire at sixty paces, a rifle, at two hundred; a chessepot, at four hundred. But I have perhaps less coolness and steadiness than at the old sixty paces, for with the rapidity of fire the new weapon is more terrible at four hundred paces, for me as well as for the enemy, than was the musket at sixty paces. And is there even more fire accuracy? No. Rifles were used before the French revolution, and yet this perfectly well known weapon was very rarely seen in war, and its efficacy, as shown in those rare cases, was unsatisfactory. Accurate fire with it at combat distances of from two hundred to four hundred meters was illusory, and it was abandoned in favor of the old rifle. Did the foot chasseurs know fire at command? Picked troops, dependable, did they use it? Yet it would have been a fine method of employing their weapons. To-day we have weapons that are accurate at six hundred to seven hundred meters. Does that mean that accurate fire at seven hundred meters is possible? No. If your enemy is armed as we are, fire at seven hundred meters will show the same results that have been shown for four hundred meters. The same losses will be suffered, and the coolness shown will be the same—that is, it will be absent. If one fire three times as fast, three times as many men will fall, and it will be three times as difficult to preserve coolness. Just as formerly it was impossible to execute fire at command, so it is to-day. Formerly no sight-setting was possible; it is no better to-day.

But if this fire is impossible, why attempt it? Let us remain always in the realm of the possible or we shall make sad mistakes. "In our art," said General Daine, "theorists abound; practical men are very rare. Also when the moment of action arrives, principles are often found to be confused, application impossible, and the most erudite officers remain inactive, unable to use the scientific treasures that they have amassed."

Let us then, practical men, seek for possible methods. Let us gather carefully the lessons of their experience, remembering Bacon's saying, "Experience excels science."

Appendix II

HISTORICAL DOCUMENTS

1. Cavalry

An Extract from Xenophon.
"The unexpectedness of an event accentuates it, be it pleasant or terrible. This is nowhere seen better than in war, where surprise terrorizes even the strongest.

"When two armies are in touch or merely separated by the field of battle, there are first, on the part of the cavalry, skirmishes, thrusts, wheels to stop or pursue the enemy, after which usually each goes cautiously and does not put forth its greatest effort until the critical part of the conflict. Or, having commenced as usual, the opposite is done and one moves swiftly, after the wheel, either to flee or to pursue. This is the method by which one can, with the least possible risk, most harm the enemy, charging at top speed when supported, or fleeing at the same speed to escape the enemy. If it is possible in these skirmishes to leave behind, formed in column and unobserved four or five of the bravest and best mounted men in each troop they may be very well employed to fall on the enemy at the moment of the wheel."

2. Marius Against the Cimbrians

Extract from Plutarch's "Life of Marius."
"Boiorix, king of the Cimbrians, at the head of a small troop of cavalry, approached Marius' camp and challenged him to fix a day and place to decide who would rule the country. Marius answered that Romans did not ask their enemies when to fight, but that he was willing to satisfy the Cimbrians. They agreed then to give battle in three days on the plain of Verceil, a convenient place for the Romans to deploy their cavalry and for the barbarians to extend their large army. The two opponents on the day set were in battle formation. Catulus had twenty thousand three hundred men. Marius had thirty-two thousand, placed on the wings and consequently on either side of those of Catulus, in the center. So writes Sylla, who was there. They say that Marius gave this disposition to the two parts of his army because he hoped to fall with his two wings on the barbarian phalanxes and wished the victory to come only to his command, without Catulus taking any part or even meeting with the enemy. Indeed, as the front of battle was very broad, the wings were separated from the center, which was broken through. They add that Catulus reported this disposition in the explanation that he had to make and complained bitterly of Marius' bad faith. The Cimbrian infantry came out of its positions in good order and in battle array formed a solid phalanx as broad as it was wide, thirty stades or about eighteen thousand feet. Their fifteen thousand horsemen were

magnificently equipped. Their helmets were crowned by the gaping mouths of savage beasts, above which were high plumes which looked like wings. This accentuated their height. They were protected by iron cuirasses and had shields of an astonishing whiteness. Each had two javelins to throw from a distance, and in close fighting they used a long heavy sword.

"In this battle the cavalry did not attack the Romans in front, but, turning to the right they gradually extended with the idea of enclosing the Romans before their infantry and themselves. The Roman generals instantly perceived the ruse. But they were not able to restrain their men, one of whom, shouting that the enemy was flying, led all the others to pursue. Meanwhile the barbarian infantry advanced like the waves of a great sea.

"Marius washed his hands, raised them to heaven, and vowed to offer a hecatomb to the gods. Catulus for his part, also raised his hands to heaven and promised to consecrate the fortune of the day. Marius also made a sacrifice, and, when the priest showed him the victim's entrails, cried, 'Victory is mine.' But, as the two armies were set in motion, something happened, which, according to Sylla, seemed divine vengeance on Marius. The movements of such a prodigious multitude raised such a cloud of dust that the two armies could not see each other. Marius, who had advanced first with his troops to fall on the enemy's formation, missed it in the dust, and having passed beyond it, wandered for a long time in the plain. Meanwhile fortune turned the barbarians toward Catulus who had to meet their whole attack with his soldiers, among whom was Sylla. The heat of the day and the burning rays of the sun, which was in the eyes of the Cimbrians, helped the Romans. The barbarians, reared in cold wooded places, hardened to extreme cold, could not stand the heat. Sweating, panting, they shaded their faces from the sun with their shields. The battle occurred after the summer solstice, three days before the new moon of the month of August, then called Sextilis. The cloud of dust sustained the Romans' courage by concealing the number of the enemy. Each battalion advancing against the enemy in front of them were engaged, before the sight of such a great horde of barbarians could shake them. Furthermore, hardship and hard work had so toughened them that in spite of the heat and impetuousness with which they attacked, no Roman was seen to sweat or pant. This, it is said, is testified to by Catulus himself in eulogizing the conduct of his troops.

"Most of the enemy, above all the bravest, were cut to pieces, for, to keep the front ranks from breaking, they were tied together by long chains attached to their belts. The victors pursued the fugitives to their entrenched camp.

"The Romans took more than sixty thousand Cimbrians prisoners, and killed twice as many."

3. The Battle of the Alma

Extract from the correspondence of Colonel Ardant du Picq. A letter sent from Huy, February 9, 1869, by Captain de V——, a company officer in the attack division.

"My company, with the 3rd, commanded by Captain D—— was designated to cover the battalion.

"At eight or nine hundred meters from the Alma, we saw a sort of wall, crowned with white, whose use we could not understand. Then, at not more than three hundred meters, this wall delivered against us a lively battalion fire and deployed at the run. It was a Russian battalion whose uniform, partridge-gray or chestnut-gray color, with white helmet, had, with the help of a bright sun, produced the illusion. This, parenthetically, showed me that this color is certainly the most sensible, as it can cause such errors. [51] We replied actively, but there was effect on neither side because the men fired too fast and too high.... The advance was then taken up, and I don't know from whom the order can have come.... We went on the run, crossing the river easily enough, and while we were assembling to scramble up the hill we saw the rest of the battalion attacking, without order, companies mixed up, crying, 'Forward,' singing, etc. We did the same, again took up the attack, and were lucky enough to reach the summit of the plateau first. The Russians, astounded, massed in a square. Why? I suppose that, turned on the left, attacked in the center, they thought themselves surrounded, and took this strange formation. At this moment a most inopportune bugle call was sounded by order of Major De M—— commanding temporarily a battalion of foot chasseurs. This officer had perceived the Russian cavalry in motion and believed that its object was to charge us, while, on the contrary it was maneuvering to escape the shells fired into it while in squadron formation by the Megere, a vessel of the fleet. This order given by bugle signal was executed as rapidly as had been the attack, such is the instinct of self-preservation which urges man to flee danger, above all when ordered to flee. Happily a level-headed officer, Captain Daguerre, seeing the gross mistake, commanded 'Forward' in a stentorian tone. This halted the retreat and caused us again to take up the attack. The attack made us masters of the telegraph-line, and the battle was won. At this second charge the Russians gave, turned, and hardly any of them were wounded with the bayonet. So then a major commanding a battalion, without orders, sounds a bugle call and endangers success. A simple Captain commands 'Forward,' and decides the victory. This is the history of yesterday, which may be useful tomorrow."

It appears from this that, apart from the able conception of the commander-in-chief, the detail of execution was abominable, and that to base on successes new rules of battle would lead to lamentable errors. Let us sum up:

First: A private chasseur d'Afrique gave the order to attack;

Second: The troops went to the attack mixed up with each other. We needed nearly an hour merely to reform the brigade. This one called, that one congratulated himself, the superior officers cried out, etc., etc.; there was confusion that would have meant disaster if the cavalry charge which was believed to threaten us, had been executed. Disorder broke out in the companies at the first shot. Once engaged, commanders of organizations no longer had them in hand, and they intermingled, so that it was not easy to locate oneself;

Third: There was no silence in ranks. Officers, non-commissioned officers and soldiers commanded, shouted, etc.; the bugles sounded the commands they heard coming from nobody knew where;

Fourth: There was no maneuvering from the first shot to the last. I do not remember being among my own men; it was only at the end that we found each other. Zouaves, chasseurs, soldiers of the 20th line formed an attack group—that was all. About four o'clock there was a first roll call. About a third of the battalion was missing at nine at night there was a second roll call. Only about fifty men were missing, thirty of whom were wounded. Where the rest were I do not know.

Fifth: To lighten the men, packs had been left on the plain at the moment fire opened, and as the operation had not been worked out in advance, no measures were taken to guard them. In the evening most of the men found their packs incomplete, lacking all the little indispensables that one cannot get in the position in which we were.

It is evidently a vital necessity to restrain the individual initiative of subordinates and leave command to the chiefs, and above all to watch the training of the soldiers who are always ready, as they approach, to run on the enemy with the bayonet. I have always noted that if a body which is charged does not hold firm, it breaks and takes flight, but that if it holds well, the charging body halts some paces away before it strikes. I shall tell you something notable that I saw at Castel-Fidardo. They talk a lot of the bayonet. For my part I only saw it used once, in the night, in a trench. Also it is noted that in the hospital, practically all the wounds treated were from fire, rarely from the bayonet.

4. The Battle of the Alma

Extract from the correspondence of Colonel A. du Picq. Letters dated in November, 1868, and February, 1869, sent from Rennes by Captain P—— of the 17th battalion of foot chasseurs, with remarks by the colonel and responses of Captain P——.

First letter from Captain P——

"... It is there that I had time to admire the coolness of my brave Captain Daguerre, advancing on a mare under the enemy's eyes, and observing imperturbable, like a tourist, all the movements of our opponents.

"I will always pay homage to his calm and collected bravery...."

Remarks by the colonel.

"Did not Captain Daguerre change the bugle call 'Retreat,' ordered by —— to the bugle call 'Forward?'"

Answer of Captain P——

"In fact, when protected in the wood by pieces of wall we were firing on the Russians, we heard behind us the bugle sounding 'Retreat' at the order of ——. At this moment my captain, indignant, ordered 'Forward' sounded to reestablish confidence which had been shaken by the distraction or by the inadvertance of ———."

5. The Battle of Inkermann

Extract from the correspondence of Colonel Ardant du Picq.
First: Letter sent from Lyon, March 21, 1869, by Major de G———, 17th Line Regiment.
"... The 1st Battalion of the 7th Light Regiment had hardly arrived close to the telegraph when it received a new order to rush to the help of the English army, which, too weak to hold such a large army, had been broken in the center of its line and driven back on its camps.

"The 1st Battalion of the 7th Light Regiment, Major Vaissier, had the honor to arrive first in the presence of the Russians, after moving three kilometers on the run. Received by the enthusiastic cheers of the English, it formed for battle, then carried away by burning cries of 'Forward, with the bayonet' from its brave major it threw itself headlong, on the Russian columns, which broke.

"For two hours the 1st Battalion of the 7th Light Regiment, a battalion of the 6th Line Regiment, four companies of the 3rd Battalion of foot chasseurs, five companies of Algerian chasseurs held the head of the Russian army which continued to debouch in massed columns from the ravine and plateau of Inkermann.

"Three times the battalion of the 7th Light Regiment was obliged to fall back some paces to rally. Three times it charged with the bayonet, with the same ardor and success.

"At four in the afternoon the Russians were in rout, and were pursued into the valley of Inkermann.

"On this memorable day all the officers, non-commissioned officers and soldiers of the 7th Light Regiment performed their duty nobly, rivalling each other in bravery and self-sacrifice."

Second: Notes on Inkermann, which Colonel A. du Picq indicates come from the letters of Captain B——— (these letters are missing).

"In what formation were the Russians? In column, of which the head fired, and whose platoons tried to get from behind the mead to enter into action?

"When Major Vaissier advanced was he followed by every one? At what distance? In what formation were the attackers? in disordered masses? in one rank? in two? in mass? Did the Russians immediately turn tail, receiving shots and the bayonet in the back? did they fall back on the mass which itself was coming up? What was the duration of this attack against a mass, whose depth prevented its falling back?

"Did we receive bayonet wounds?

"Did we fall back before the active reaction of the mass or merely because, after the first shock, the isolated soldiers fell back to find companions and with them a new confidence?

"Was the second charge made like the first one? Was the 6th Line Regiment engaged as the first support of the 7th Light Regiment? How were the Zouaves engaged?"

6. The Battle of Magenta

Extract from the correspondence of Colonel Ardant du Picq. Letters from Captain C——, dated August 23, 1868.

"At Magenta I was in Espinasse's division, of Marshal MacMahon's corps. This division was on the extreme left of the troops that had passed the Ticino at Turbigo and was moving on Magenta by the left bank. Close to the village a fusillade at close range apprised us that the enemy was before us. The country, covered with trees, hedges, and vines, had hidden them.

"Our 1st Battalion and the 2nd Foreign Regiment drove the Austrians into Magenta.

"Meanwhile the 2nd and 3rd Battalions of Zouaves, with which I was, remained in reserve, arms stacked, under control of the division commander. Apparently quite an interval had been left between Espinasse's division and la Motterouge's, the 1st of the corps, and, at the moment of engagement, at least an Austrian brigade had entered the gap, and had taken in flank and rear the elements of our division engaged before Magenta. Happily the wooded country concealed the situation or I doubt whether our troops engaged would have held on as they did. At any rate the two reserve battalions had not moved. The fusillade extended to our right and left as if to surround us; bullets already came from our right flank. The General had put five guns in front of us, to fire on the village, and at the same time I received the order to move my section to the right, to drive off the invisible enemy who was firing on us. I remember that I had quit the column with my section when I saw a frightened artillery captain run toward us, crying 'General, General, we are losing a piece!' The general answered, 'Come! Zouaves, packs off.' At these words, the two battalions leaped forward like a flock of sheep, dropping packs everywhere. The Austrians were not seen at first. It was only after advancing for an instant that they were seen. They were already dragging off the piece that they had taken. At the sight of them our men gave a yell and fell on them. Surprise and terror so possessed the Austrians, who did not know that we were so near, that they ran without using their arms. The piece was retaken; the regimental standard was captured by a man in my company. About two hundred prisoners were taken, and the Austrian regiment—Hartmann's 9th Infantry—was dispersed like sheep in flight, five battalions of them. I believe that had the country not been thick the result might have been different. The incident lasted perhaps ten minutes.

"The two battalions took up their first position. They had had no losses, and their morale was in the clouds. After about an hour General Espinasse put himself at the head of the two battalions and marched us on the village. We were in column of platoons with section intervals. The advance was made by echelon, the 2nd Battalion in front, the 3rd a little in rear, and a company in front deployed as skirmishers.

"At one hundred and fifty paces from the Austrians, wavering was evident in their lines; the first ranks threw themselves back on those in rear. At that instant the general ordered again, 'Come! Packs off. At the double!' Everybody ran forward, shedding his pack where he was.

"The Austrians did not wait for us. We entered the village mixed up with them. The fighting in houses lasted quite a while. Most of the Austrians retired. Those who remained in the houses had to surrender. I found myself, with some fifty officers and men, in a big house from which we took four hundred men and five officers, Colonel Hauser for one.

"My opinion is that we were very lucky at Magenta. The thick country in which we fought, favored us in hiding our inferior number from the Austrians. I do not believe we would have succeeded so well in open country. In the gun episode the Austrians were surprised, stunned. Those whom we took kept their arms in their hands, without either abandoning them or using them. It was a typical Zouave attack, which, when it succeeds, has astonishing results; but if one is not lucky it sometimes costs dearly. Note the 3rd Zouaves at Palestro, the 1st Zouaves at Marignano. General Espinasse's advance on the village, at the head of two battalions, was the finest and most imposing sight I have ever seen. Apart from that advance, the fighting was always by skirmishers and in large groups."

7. The Battle of Solferino

Extract from the correspondence of Colonel Ardant du Picq. Letters from Captain C——.

"The 55th infantry was part of the 3rd division of the 4th corps.

"Coming out of Medole, the regiment was halted on the right of the road and formed, as each company arrived, in close column. Fascines were made.

"An aide-de-camp came up and gave an order to the Colonel.

"The regiment was then put on the road, marched some yards and formed in battalion masses on the right of the line of battle. This movement was executed very regularly although bullets commenced to find us. Arms were rested, and we stayed there, exposed to fire, without doing anything, not even sending out a skirmisher. For that matter, during the whole campaign, it seemed to me that the skirmisher school might never have existed.

"Then up came a Major of Engineers, from General Niel, to get a battalion from the regiment. The 3rd battalion being on the left received the order to march. The major commanding ordered 'by the left flank,' and we marched by the flank, in close column, in the face of the enemy, up to Casa-Nova Farm, I believe, where General Niel was.

"The battalion halted a moment, faced to the front, and closed a little.

"'Stay here,' said General Niel; 'you are my only reserve!'

"Then the general, glancing in front of the farm, said to the major, after one or two minutes, 'Major, fix bayonets, sound the charge, and forward!'

"This last movement was still properly executed at the start, and for about one hundred yards of advance.

"Shrapnel annoyed the battalion, and the men shouldered arms to march better.

"At about one hundred yards from the farm, the cry 'Packs down,' came from I do not know where. The cry was instantly repeated in the battalion. Packs were thrown down, anywhere, and with wild yells the advance was renewed, in the wildest disorder.

"From that moment, and for the rest of the day, the 3rd Battalion as a unit disappeared.

"Toward the end of the day, after an attempt had been made to get the regiment together, and at the end of half an hour of backing and filling, there was a roll-call.

"The third company of grenadiers had on starting off in the morning one hundred and thirty-two to one hundred and thirty-five present. At this first roll-call, forty-seven answered, a number I can swear to, but many of the men were still hunting packs and rations. The next day at reveille roll-call, ninety-three or four answered. Many came back in the night.

"This was the strength for many days I still remember, for I was charged with company supply from June 25th.

"As additional bit of information—it was generally known a few days later that at least twenty men of the 4th company of grenadiers were never on the field of battle. Wounded of the company, returned for transport to Medole, said later that they had seen some twenty of the company together close to Medole, lying in the grass while their comrades fought. They even gave some names, but could not name them all. The company had only been formed for the war on April 19th, and had received that same day forty-nine new grenadiers and twenty-nine at Milan, which made seventy-eight recruits in two months. None of these men were tried or punished. Their comrades rode them hard, that was all."

8. Mentana

Extract from the correspondence of Colonel Ardant du Picq. Letters from Captain C——, dated August 23, 1868.

"November 3, at two in the morning, we took up arms to go to Monte-Rotondo. We did not yet know that we would meet the Garibaldians at Mentana.

"The Papal army had about three thousand men, we about two thousand five hundred. At one o'clock the Papal forces met their enemies. The Zouaves attacked vigorously, but the first engagements were without great losses on either side. There is nothing particular in this first episode. The usual thing happened, a force advances and is not halted by the fire of its adversary who ends by showing his heels. The papal Zouaves are marked by no ordinary spirit.

In comparing them with the soldiers of the Antibes legion, one is forced to the conclusion that the man who fights for an idea fights better than one who fights for money. At each advance of the papal forces, we advanced also. We were not greatly concerned about the fight, we hardly thought that we would have to participate, not dreaming that we could be held by the volunteers. However, that did not happen.

"It was about three o'clock. At that time three companies of the battalion were employed in protecting the artillery—three or four pieces placed about the battle-field. The head of the French column was then formed by the last three companies of the battalion, one of the 1st Line Regiment; the other regiments were immediately behind. Colonel Fremont of the 1st Line Regiment, after having studied the battle-field, took two chasseur companies, followed by a battalion of his regiment and bore to the right to turn the village.

"Meanwhile the 1st Line Regiment moved further to the right in the direction of Monte-Rotondo, against which at two different times it opened a fire at will which seemed a veritable hurricane. Due to the distance or to the terrain the material result of the fire seemed to be negligible. The moral result must have been considerable, it precipitated a flood of fugitives on the road from Mentana to Monte-Rotondo, dominated by our sharpshooters, who opened on the fugitives a fire more deadly than that of the chassepots. We stayed in the same position until night, when we retired to a position near Mentana, where we bivouacked.

"My company was one of the two chasseur companies which attacked on the right with the 1st Line Regiment. My company had ninety-eight rifles (we had not yet received the chassepots). It forced the volunteers from solidly held positions where they left a gun and a considerable number of rifles. In addition, it put nearly seventy men out of action, judging by those who remained on the field. It had one man slightly wounded, a belt and a carbine broken by bullets.

"There remained with the general, after our movement to the right, three companies of chasseurs, a battalion of the 29th, and three of the 59th. I do not include many elements of the Papal army which had not been engaged. Some of my comrades told me of having been engaged with a chasseur company of the 59th in a sunken road, whose sides had not been occupied; the general was with this column. Having arrived close to the village, some shots either from the houses or from enemy sharpshooters, who might easily have gotten on the undefended flanks, provoked a terrible fusillade in the column. In spite of the orders and efforts of the officers, everybody fired, at the risk of killing each other, and this probably happened. It was only when some men, led by officers, were able to climb the sides of the road that this firing ceased. I do not think that this was a well understood use of new arms.

"The fusillade of the 1st Line Regiment against Monte-Rotondo was not very effective, I believe negligible. I do not refer to the moral result, which was great.

"The Garibaldians were numerous about Monte-Rotondo. But the terrain like all that around Italian villages was covered with trees, hedges, etc. Under these conditions, I believe that the fire of sharpshooters would have been more effective than volleys, where the men estimate distances badly and do not aim."

NOTES

[1] General Daumas (Manners and Customs of Algeria). Nocturnal Surprise and Extermination of a Camp.

[2] Among the Romans, mechanics and morale are so admirably united, that the one always comes to the aid of the other and never injures it.

[3] The Romans did not make light of the influence of a poet like Tyrtaeus. They did not despise any effective means. But they knew the value of each.

[4] Also their common sense led them to recognize immediately and appropriate arms better than their own.

[5] This is an excuse. The maniple was of perfect nobility and, without the least difficulty, could face in any direction.

[6] This was an enveloping attack of an army and not of men or groups. The Roman army formed a wedge and was attacked at the point and sides of the wedge; there was not a separate flank attack. That very day the maniple presented more depth than front.

[7] They had been sent to attack Hannibal's camp; they were repulsed and taken prisoner in their own camp after the battle.

[8] This extract is taken from the translation of Dom Thuillier. Livy does not state the precise number of Roman combatants. He says nothing had been neglected in order to render the Roman army the strongest possible, and from what he was told by some it numbered eighty-seven thousand two hundred men. That is the figure of Polybius. His account has killed, forty-five thousand; taken or escaped after the action, nineteen thousand. Total sixty-four thousand. What can have become of the twenty-three thousand remaining?

[9] The Numidian horsemen were a light irregular cavalry, excellent for skirmishing, harassing, terrifying, by their extraordinary shouts and their unbridled gallop. They were not able to hold out against a regular disciplined cavalry provided with bits and substantial arms. They were but a swarm of flies that always harasses and kills at the least mistake; elusive and perfect for a long pursuit and the massacre of the vanquished to whom the Numidians gave neither rest nor truce. They were like Arab cavalry, badly armed for the combat, but sufficiently armed for butchering, as results show. The Arabian knife, the Kabyle knife, the Indian knife of our days, which is the favorite of the barbarian or savage, must play its part.

[10] They formed the third Roman line according to the order of battle of the Legion. The contraction of the first line into a point would naturally hem them in.

[11] Brought back by Hannibal who had reserved to himself the command of the center.

[12] The triarians, the third Roman line.

[13] What effect this might have, was shown in the battle of Alisia, where Caesar's men, forewarned by him, were nevertheless troubled by war-whoops behind them. The din of battle in rear has always demoralized troops.

[14] His cavalry consisted of seven thousand horse, of which five hundred were Gauls or Germans, the best horsemen of that time, nine hundred Galicians, five hundred Thracians, and Thessalians, Macedonians and Italians in various numbers.

[15] Caesar's legions in battle order were in three lines: four cohorts in the first line, two in the second, and three in the third. In this way the cohorts of a legion were, in battle, always supported by cohorts of the same legion.

[16] Caesar stated that in order to make up the numerical inferiority of his cavalry, he had chosen four hundred of the most alert young men, from among those marching ahead of the standards, and by daily exercise had them accustomed to fighting between his horsemen. He had in this way obtained such results that his thousand riders dared, in open field, to cope with Pompey's seven thousand cavalry without becoming frightened at their number.

[17] Any one who wishes to read in extenso is referred to the fight of the ten thousand against Pharnabazus in Bithynia, Xenophon, par. 34, page 569, Lisken & Sauvan edition.—In Polybius, the battle of the Tecinus, Chapt. XIII, of Book III.—In Caesar or those who followed him the battles against Scipio, Labienus, and Afranius, the Getae and the Numidians, par. 61, page 282, and par. 69, 70, 71 and 72, pp. 283, 285, and 286, in the African war, Lisken & Sauvan edition.

[18] In ancient combat, there was almost only, dead or lightly wounded. In action, a severe wound or one that incapacitated a man was immediately followed by the finishing stroke.

[19] Hand-to-hand, sword-to-sword, serious fighting at short distances, was rare then. Likewise in the duels of our day blades are rarely crossed in actual practice.

[20] To-day, it is the riflemen who do nearly all the work of destruction.

[21] Considering Caesar's narrative what becomes of the mathematical theory of masses, which is still discussed? If that theory had the least use, how could Marius ever have held out against the tide of the armies of the Cimbri and Teutons? In the battle of Pharsalus, the advice given by Triarius to Pompey's army, a counsel which was followed and which was from a man of experience, who had seen things close at hand, shows that the shock, the physical impulse of the mass was a by-word. They knew what to think of it.

[22] The individual advance, in modern battle, in the midst of blind projectiles that do not choose, is much less dangerous than in ancient times, because it seldom goes up to the enemy.

At Pharsalus, the volunteer Crastinius, an old centurion, moved ahead with about a hundred men, saying to Caesar: "I am going to act, general, in such a way that, living or dead, to-day you may have cause to be proud of me."

Caesar, to whom these examples of blind devotion to his person were not displeasing, and whose troops had shown him that they were too mature, too experienced, to fear the contagion of this example, let Crastinius and his companions go out to be killed.

Such blind courage influences the action of the mass that follows. Probably for that reason, Caesar permitted it. But against reliable troops, as the example of

Crastinius proves, to move ahead in this way, against the enemy, is to go to certain death.

[23] The men of the maniple, of the Roman company, mutually gave their word never to leave ranks, except to pick up an arrow, to save a comrade (a Roman citizen), or to kill an enemy. (Livy).

[24] A small body of troops falling into a trap might present a sort of mêlée, for a second, the time necessary for its slaughter. In a rout it might be possible at some moment of the butchery to have conflict, a struggle of some men with courage, who want to sell their lives dearly. But this is not a real mêlée. Men are hemmed in, overwhelmed, but not thrown into confusion.

[25] The Greek phalanx.

[26] The Romans lost no one as their companies entered the openings in the phalanx.

[27] The Roman velites, light-armed soldiers, of the primitive legion before Marius, were required to stand for an instant in the intervals of the maniples, while awaiting the onset. They maintained, but only for an instant, the continuity of support.

[28] A result forced by the improvement of war appliances.

[29] In troops without cohesion, this movement begins at fifty leagues from the enemy. Numbers enter the hospitals without any other complaint than the lack of morale, which very quickly becomes a real disease. A Draconian discipline no longer exists; cohesion alone can replace it.

[30] It is a troublesome matter to attack men who shoot six to eight shots a minute, no matter how badly aimed. Will he have the last word then, who has the last cartridge, who knows best how to make the enemy use his cartridges without using his own?

The reasoning is always the same. With arrows: Let us use up their arrows. With the club: Let us break their clubs. But how? That is always the question. In matters of war, above all, precept is easy; accomplishment is difficult.

[31] The more one imagines he is isolated, the more has he need of morale.

[32] Are not naval battles above all the battles of captains? All captains endeavor to promote a feeling of solidarity which will cause them all to fight unitedly on the day of action. Trafalgar—Lissa.

In 1588, the Duke of Medina Sidonia, preparing for a naval engagement, sent three commanders on light vessels to the advance-guard and three to the rearguard, with executioners, and ordered them to have every captain hanged who abandoned the post that had been assigned to him for the battle.

In 1702, the English Admiral Benbow, a courageous man, was left almost alone by his captains during three days of fighting. With an amputated leg and arm, before dying, he had four brought to trial. One was acquitted, three were hanged; and from that instant dates the inflexible English severity towards commanders of fleets and vessels, a severity necessary in order to force them to fight effectively.

Our commanders of battalions, our captains, our men, once under fire, are more at sea than these commanders of vessels.

[33] The effect of surprise would certainly not last long to-day. However, to-day wars are quickly decided.
[34] See Appendix VI. (Historical documents). (Editor's note).
[35] See Appendix VI. (Historical documents). (Editor's note).
[36] See Appendix VI. (Historical documents). (Editor's note).
[37] See Appendix VI. (Historical documents). (Editor's note).
[38] See Appendix VI. (Historical documents). (Editor's note).
[39] It is true that such measures are recommended in camps of instruction and in publications. But in maneuvers they are neglected in the mania for alignment, and in that other mad desire of generals to mix in details which do not concern them.
[40] See Appendix VI. (Historical documents.) (Editor's note.)
[41] See Appendix VI. (Historical documents.) (Editor's note.)
[42] See Appendix II. (Historical documents.) (Editor's note.)
[43] A propos of gaps: At the battle of Sempach thirteen hundred badly armed Swiss opposed three thousand Lorraine knights in phalanxes. The attack of the Swiss in a formation was ineffective, and they were threatened with envelopment. But Arnold von Winkelried created a gap; the Swiss penetrated and the massacre followed.
[44] See Appendix II. (Historical documents.) (Editor's note.)
[45] See Appendix II. (Historical documents.) (Editor's note.)
[46] See Appendix II. (Historical documents.) (Editor's note.)
[47] It is hard to determine what method of fire, at command or at will, was used. But what we find in the works of the best military authorities, from Montecuculli to Marshal Saxe, is general opposition to the replacement of the pike by the rifle. All predicted the abandonment of the rifle for the pike, and the future always proved them wrong. They ignored experience. They could not understand that stronger than all logic is the instinct of man, who prefers long range to close fighting, and who, having the rifle would not let it go, but continually improved it.
[48] The danger arising from this kind of fire, led to proposals to put the smallest men in the front rank, the tallest in the rear rank.
[49] Nothing is more difficult than to estimate range; in nothing is the eye more easily deceived. Practice and the use of instruments cannot make a man infallible. At Sebastopol, for two months, a distance of one thousand to twelve hundred meters could not be determined by the rifle, due to inability to see the shots. For three months it was impossible to measure by ranging shots, although all ranges were followed through, the distance to a certain battery which was only five hundred meters away, but higher and separated from us by a ravine. One day, after three months, two shots at five hundred meters were observed in the target. This distance was estimated by everybody as over one thousand meters; it was only five hundred. The village taken and the point of observation changed, the truth became evident.
[50] His war instructions prove this. His best generals, Zieten, Warnery, knew of such methods, saw nothing practicable in them and guarded against them in war

as indeed he did himself. But Europe believed him, tried to imitate his maneuvers on the field of battle, and aligned her troops to be beaten by him. This is what he was after. He even deceived the Prussians. But they came back to sound methods after 1808, in 1813 and afterwards.

[51] It is noted here that French uniforms are of an absurd color, serving only to take the eye at a review. So the chasseurs, in black, are seen much further than a rifleman of the line in his gray coat. The red trousers are seen further than the gray—thus gray ought to be the basic color of the infantry uniform, above all that of skirmishers.

At night fall the Russians came up to our trenches without being seen by any one, thanks to their partridge-gray coats.

The Echo Library
www.echo-library.com

Please visit our website to download a complete catalogue of our books. We specialise in out-of-copyright reprints in all subjects. We publish a range of Large Print Editions of classic titles.

Books for re-printing

Suggestions for re-printing can be sent to titles@echo-library.com. Titles must be out-of-copyright or you must own the copyright. They can be in digital or published form.

If we do not wish to add the title to our catalogue then we can reprint for you. Costs depend on the format of the manuscript or book and its size.

Feedback

Because we have no direct contact with our customers we would welcome constructive comments to feedback@echo-library.com

Complaints

If there is a serious error in the text or layout please advise us by sending details to complaints@echo-library.com and we will supply a corrected copy, usually within 15 working days (it may take longer outside the UK and USA).

If there is a printing fault or the book is damaged please refer to your supplier.

Printed in the United States
75805LV00008B/118